ENCYCLOPAEDIA OF
NANO RESEARCH

ENCYCLOPAEDIA OF
NANO RESEARCH

J. Fox

ANMOL PUBLICATIONS PVT. LTD.
NEW DELHI - 110 002 (INDIA)

ANMOL PUBLICATIONS PVT. LTD.
H.O.: 4374/4B, Ansari Road, Darya Ganj,
New Delhi-110 002 (India)
Ph.: 23278000, 23261597

B.O.: No. 1015, Ist Main Road, BSK IIIrd Stage
IIIrd Phase, IIIrd Block,
Bangalore - 560 085 (India)
Visit us at: www.anmolpublications.com

Encyclopaedia of Nano Research

First Published, 2006

ISBN 81-261-2978-6

PRINTED IN INDIA

Printed at Mehra Offset Press, Delhi.

Contents

Preface

This book entitled "Encyclopaedia of Nano Research" covers extensively the major fields where research in on global in the fields of nanoscale science and technology. It also a briefly on areas where future focus lies and nano applications coming in. The cover the present opportunities in nano characterisation, manipulation and self assembly, the development principles of nanotechnology. Nano-research materials do not represent a new phenomenon. However, the ability to probe, manipulate, understand and engineer matter at atomic scales has only recently come within our grasp. Developments such as the invention of the Scanning Tunneling Microscope in 1981 have since made nanoscale science a reality. Today, the term nanotechnology is used to describe specific construction and manipulation performed on an atomic and molecular scale. Generally, systems with a maximum size of less than a hundred nanometers are classified as nanotechnology.

This is the equivalent of approximately five to ten atoms laid end-to-end. These dimensions are also reflected in the English translation of the Greek word "nanos"—"dwarf". A single human hair is around ten thousand times thicker than a nanometer, and another impressive comparison – a nanoparticle is to a soccer ball as a soccer ball is to the Earth. But nanotechnology is not just microtechnology on a smaller scale; it is much more than a mere improvement on previous forms of technology. Nanotechnology is a completely new production process in a molecular world, opening up possibilities far beyond what we have known before. Nature provides the inspiration for this new technology – from time immemorial, nanomachines have been at work in all organic cells, building our world of plants, animals and people from a store of chemical elements. The discovery in recent years of the mechanisms for accessing individual building blocks of matter, and with it the increasing understanding of how these building blocks organize themselves, has sparked off the industrial conquest of the nanosphere throughout the world. This book tends to cover the aforementioned subject areas in their broadest possible terms. All useful supplementary research and reference tools have been included for readers' further investigations in the subject area of Nano Research.

—Editor

1

Nano Research: An Interdisciplinary Understanding

There are some outstanding opportunities and challenges that face the nanoscience community. For dispersions and coatings these include four areas:

1. The foremost area of opportunity is controlling the particle preparation process so that size is reproducible and scaleable. This requires creation of narrow size range particles that can be prepared by processes mentioned in the study such as vapor phase condensation, physical size reduction, or flame and pyrolysis aerosol generators. These same processes must respond to good reproducibility and scaling. In most studies to date, the size of primary particles depends on material properties and the temperature/time history. Two processes, collision and coalescence, occur together, and the processes need to be controlled in order to favorably influence final particle size distribution (Wu et al. 1993).

2. The second area of opportunity is process control (Henderson 1996). In the concept of a process control methodology, the nature of chemical processes makes it imperative to have means of effectively monitoring and initiating change in the process variables of interest. Accordingly, those involved with production of nanoparticles would monitor outputs, make decisions about how to manipulate outputs in order to obtain desired behaviors, and then implement these decisions on the process. Control system configuration will necessarily have a feeding process from the output such that information can be fed back to the controller. It may also have an opportunity to base controller decision-making on information that is being fed forward; decisions could be made before the process is affected by incoming disturbances. Such would apply to process parameters like temperature, flow rate, and pressure. In many chemical setups, product quality variables must be considered, and such measurements often take place in the laboratory. An objective analysis is needed of any observed deviation of a process variable from its aim, and an objective decision must be made as to what must be done to minimize any deviation. Here process understanding leads to well

behaved reactors in manufacturing that allow for true process verification with data feedback and analysis. This concept of process control requires the application of statistical methodologies for product and process improvements. Current interest in statistical process control (SPC) is due to several factors, in particular, interest in realizing consistent high quality so as to sustain the business and obtain greater market share. Most manufacturing zones are moving toward inline/online sensor technologies coupled with process software and user-friendly computer hardware for factory operators. The benefits of process control are many. They include achieving reduced variability and higher quality, safety enhancement, reduction of process upsets, and in many cases, environmental improvements due to achieving mass balance in processes with material in/product out. Poor process design can be inherently overcome through SPC. Reduction in sampling and inspection costs results.

3. A third area of opportunity is the process/product relationship that leads to continuous uniformity—that is, the specification setting by product users that must be available to relate to process control in manufacturing. Here, nanoparticle formulation and the process used to prepare the particles must be linked and interactive. 4. A fourth technical opportunity is to develop process models for various dispersions and coatings that lead to shorter cycle time in manufacturing. A comparison of enablers and opportunities in Europe, Japan, and the United States. These four areas of opportunity, then, represent what lies ahead for advancement in nanotechnology with respect to dispersions and coatings. Achieving implementation in the industrial market will require close attention to resolving the challenges in the four areas summarized above. In addition, there remains a large gap between the cost of preparing conventional materials and the cost of preparing nanoparticles. This will remain a challenge for the future if nanomaterials are to be competitive.

Nanostructure

The field of nanostructure science and technology is a broad and interdisciplinary area of worldwide research and development activity that has been growing explosively in the past few years. While an understanding of the range and nature of functionalities that can be accessed through nanostructuring is just beginning to unfold, its tremendous potential for revolutionizing the ways in which materials and products are created is already clear. It is already having a significant commercial impact, and it will very certainly have a much greater impact in the future. During the years 1996-98, an eight-person panel under the auspices of the World Technology Evaluation Center (WTEC) conducted a worldwide study of the research and development status and trends in nanoparticles, nanostructured materials, and nanodevices, or more concisely, nanostructure science and technology. The study was commissioned and sponsored by a wide range of U.S. government agencies led by the National Science Foundation (NSF), which included the Air Force Office of Scientific Research, Office of Naval Research, Department of Commerce (including the National Institute of Standards and Technology and the Technology Administration), Department of Energy, National Institutes of Health, and National Aeronautics and Space Administration.

Additional participating U.S. government agencies for the study were the Army Research Office, Army Research Laboratory, Defense Advanced Research Projects Agency, and the Ballistic Missile Defense Organization. The uniquely broad sponsor list for this WTEC study mirrors the broadly based interests in and, in fact, the reality of the field of nanostructure science and technology. The panel study began in 1996, when panel co-chair Prof. Evelyn L. Hu (University of California at Santa Barbara) and I came to Washington to present our thoughts to WTEC and the sponsors on how the study could best be configured and carried out, given the available resources (time, people, and money).

After an extensive discussion with sponsors and potential sponsors of the study, we assembled a team of experts for the panel from industry and university, including Dr. Donald M. Cox (Exxon Research and Engineering Company), Dr. Herb Goronkin (Motorola), Prof. Lynn Jelinski (Cornell University during most of this study, now at Louisiana State University), Prof. Carl Koch (North Carolina State University), John Mendel (Eastman Kodak Company), and Prof. David T. Shaw (State University of New York at Buffalo). Two of us on the panel, Prof. Koch and I, although presently in universities, had spent large fractions of our careers at Oak Ridge and Argonne National Laboratories, respectively, lending national laboratory perspectives to the study, as well.

The purposes of this study, which the panel determined in conjunction with its sponsors, were to assess the current status and future trends internationally in research and development in the broad and rapidly growing area of nanostructure science and technology. The study had the following four goals: 1. to provide the worldwide science and engineering community with a broadly inclusive and critical view of this field 2. to identify promising areas for future research and commercial development 3. to help stimulate development of an interdisciplinary international community of nanostructure researchers 4. to encourage and identify opportunities for international collaboration Based on these goals, the panel formulated a number of questions, for which we sought answers during our study: • What are the scientific drivers (new properties and phenomena, instruments, theory, and simulation methods) and advantages (applications) to be gained from control at the nanostructure level? • What are the critical parameters to control in nanostructured material synthesis and device manufacturing? • What are the likelihood of and the time scale for bringing these new technologies to fruition?

What are the underlying research and development and educational concepts and directions driving nanostructure science and technology development? What are the expected financial dimensions of this effort over the next five to ten years, and are there national programs in force or planned? Which areas of nanostructure science and technology would be most fruitful for international collaboration? These questions were posed in advance to the various workshop participants and hosts of our panel visits so that answers could be considered and prepared. In every case, our hosts went to great lengths and considerable efforts to prepare for our visits and to make our study both effective and comfortable.

The various activities of the WTEC panel, in addition to considerable reading, thinking, discussing, and writing, included the following: (i) a U.S. workshop on 8-9 May 1997 with presentations by 26 invited expert participants from universities, industry, and national laboratories, and by 23 U.S. government agency sponsors (ii) visits by panel members to 42 universities, industrial companies, and national laboratories in Europe (France, Germany, Belgium, the Netherlands, Sweden, Switzerland, and the United Kingdom), Japan, and Taiwan (iii) three round-table workshops involving 27 additional institutions in Germany, Sweden, and Russia These activities represent a rather broad base of information for the study from which the panel derived the findings and conclusions that appear in this volume. One must emphasize, however, that even though we visited many places, listened to many presentations, and read much material, there is no way that this study is, or could have been with the available resources, encyclopedic.

The field of nanostructure science and technology is simply too large, too geographically dispersed, and changing too rapidly to cover exhaustively. What this volume presents are only examples, the best examples the panel could find, to describe what the field encompasses, its current breath and depth, and where it appears to be heading. The choices of the places that we visited, and even the types of visits, were made from lists and suggestions generated by all the panel members, with useful sponsor input. The final priorities were made according to where we felt the most exciting research and development activities in nanostructure science and technology were going on, overlayed with a realistic evaluation of which sites and how many of them could be logically visited in the time allotted.

Unfortunately, this means that panel members on our limited schedule could not visit many interesting institutions and could not accommodate visits to entire countries that make significant contributions to nanostructure science and technology—Australia, Canada, China, Finland, India, Israel, Italy, Mexico, Spain, and the Ukraine, to name just a few. The initial public report of the findings of the WTEC study panel was presented in Arlington, Virginia (http://itri.loyola.edu/nano/views/top.htm) on 10 February 1998. Full site reports from the panel's visits in Europe, Japan, and Taiwan are included as Appendices B to E in this volume. A separate volume covering the U.S. workshop has already been published by WTEC under the title *R&D Status and Trends in Nanoparticles, Nanostructured Materials, and Nanodevices in the United States* .

WTEC will soon publish a third volume of this study that consists of papers presented at its workshop in St. Petersburg, Russia. An introduction to and overview of the study and its conclusions are presented in this chapter, including some of the technical highlights of nanostructure science and technology that the panel observed. The WTEC panel would like to take this opportunity to thank all of the study participants around the world for their conscientious help and contributions and for their generous hospitality. We would also like to extend our thanks to Dr. Mike Roco of NSF for the wonderful support he has given us throughout the study and for his active participation in many of the visits that we made around the world. In addition, we would like to thank Mr. Geoff Holdridge (WTEC Director) and his staff for their excellent support, without which the study could not have been accomplished.

There are two overarching findings from this WTEC study. First, it is now abundantly clear that we are able to nanostructure materials for novel performance. This is the essential theme of this field: novel performance through nanostructuring. Nanostructuring represents the beginning of a revolutionary new age in our ability to manipulate materials for the good of humanity. The synthesis and control of materials in nanometer dimensions can access new material properties and device characteristics in unprecedented ways.

Panelists had seen the tip of the iceberg or the pinnacle of the pyramid before starting this study, but only since undertaking the study do we fully appreciate just how broad the field really is and begin to understand what its exciting potential and impact may really be. It is now clear that work is rapidly expanding worldwide in exploiting the opportunities offered through nanostructuring. The second major finding is that there is a wide range of disciplines contributing to the developments in nanostructure science and technology worldwide. Each year sees an ever increasing number of researchers from diverse disciplines enter the field and an increasing breadth of novel ideas and exciting new opportunities explode on the international nanostructure scene.

The rapidly mounting level of interdisciplinary activity in nanostructuring is truly exciting. The intersections between the various disciplines are where much of the novel activity resides, and this activity is growing in importance. If nothing else, these are the two basic findings that you need to carry away from this study. "building blocks" atoms nanoparticles layers nanostructures dispersions and coatings high surface area materials functional nanodevices consolidated materials synthesis assembly. Organization of nanostructure science and technology and the WTEC study.

The basis of the field is any type of material (metal, ceramic, polymer, semiconductor, glass, composite) created from nanoscale building blocks (clusters or nanoparticles, nanotubes, nanolayers, etc.) that are themselves synthesized from atoms and molecules. Thus, the controlled synthesis of those building blocks and their subsequent assembly into nanostructures is one fundamental theme of this field. This is the subject of this chapter by Evelyn Hu and David Shaw. This theme draws upon all of the materialsrelated disciplines from physics to chemistry to biology and to essentially all of the engineering disciplines as well. In fact, there is a very strong thread from all these disciplines running throughout the fabric of this study; the biological aspects are so pervasive that special attention is given to them. The second and most fundamentally important theme of this field is that the nanoscale building blocks, because of their sizes below about 100 nm, impart to the nanostructures created from them new and improved properties and functionalities heretofore unavailable in conventional materials and devices.

The reason for this is that materials in this size range can exhibit fundamentally new behavior when their sizes fall below the critical length scale associated with any given property. Thus, essentially any material property can be dramatically changed and engineered through the controlled size-selective synthesis and assembly of nanoscale building blocks. Four broadly defined and overlapping application areas that cover the tremendous range of challenges and opportunities for nanostructure science and technology are dispersions and coatings, high surface area materials, functional nanodevices, and consolidated materials. In the synthesis and assembly area we see that atoms, molecules, clusters and nanoparticles can be used as building blocks for nanostructuring.

However, the useful size of these building blocks depends upon the property to be engineered, since the critical length scales for which one is designing these building blocks depends upon the particular property of interest. For multifunctional applications, more than one property and one length scale must be considered. Every property has a critical length scale, and if a nanoscale building block is made smaller than that critical length scale, the fundamental physics of that property starts to change. By altering the sizes of those building blocks, controlling their internal and surface chemistry, and controlling their assembly, it is possible to engineer properties and functionalities in unprecedented ways. The characteristics of the building blocks, such as their size and size distribution, composition, composition variation, and morphology, must be well controlled.

Also, the interfaces between the building blocks and their surroundings can be critical to performance. It is not sufficient simply to make the building blocks; one must also worry about the structure and chemistry of their surfaces and how they will interact one with another or with a matrix in which they are embedded. There is a very wide range of diverse synthesis and assembly strategies being employed in nanostructuring, all the way from fundamental biological methods for selfassembling molecules to sophisticated chemical precipitation methods to a variety of physical and chemical aerosol techniques for making clusters or nanoparticles and then dispersing them or bringing them together in consolidated forms. All of these strategies contribute in essential ways to the growth of this field. Each may have unique capabilities that will benefit a particular property, application, or process. The most generally applicable of them are likely to have significant technological impact and commercial potential.

In the area of dispersions and coatings, a wide range of new and enhanced functionalities are now becoming available by means of nanostructuring. They cover the whole set of properties that are of interest in optical, thermal, and electrical applications. This is the most mature area of nanoscale science and technology. The many current commercial applications include printing, sunscreens, photography, and pharmaceuticals. Some examples of the present technological impact of nanostructuring are thermal and optical barriers, imaging enhancement, ink-jet materials, coated abrasive slurries, and information-recording layers. From our vantage point at present, there appears to be very strong potential impact in the areas of targeted drug delivery, gene therapy, and multifunctional coatings. Nevertheless, certain central issues must be addressed if work in this area is going to continue to affect society in meaningful new ways in the coming years.

Successful nanoscale dispersions require freedom from agglomeration and surface control. Process controls are required to ensure reproducibility, reliability, and scalability. There is also a need to develop process models that lead to shorter cycle times in manufacturing, if commercialization is to be truly effective. In the area of high surface area materials, reviewed by Donald Cox, it is of primary importance to realize that nanostructured material building blocks have inherently high surface areas unless they are consolidated. For example, a nanoparticle 5 nm in diameter has about half of its atoms on its surface. If the nanoparticles are then brought together in a lightly assembled way, this surface area is available for a variety of useful applications.

In fact, there is a wide range of new applications in high capacity uses for chemical and electrical energy storage, or in sensors and other applications that take copious advantage of this feature. Already there are numerous commercial applications in porous membranes or molecular sieves, drug delivery, tailored catalysts, and absorption/desorption materials. Clearly, what is required to optimize the impact of nanostructures to be really useful to society in high surface area material applications is to create materials that combine high selectivity, high product or function yield, and high stability. Thus, the major challenges in this area are critical dimensional control and long-term thermal and chemical stability.

When these problems are solved, considerable future technological potential is seen in the areas of molecule-specific sensors, large hydrocarbon or bacterial filters, energy storage, and Grätzel-type solar cells. The area of functional nanoscale devices, covered in this chapter by Herb Goronkin and his Motorola colleagues, is largely driven by the need for ever smaller devices, which necessitate both new device and new circuit architectures. It is not very useful to make nanoscale devices if they cannot be assembled in a circuit with interconnects that are themselves nanoscale. Thus, a complete rethinking of this area is required. The major research and development effort worldwide in functional nanoscale devices is focused on the single electron transistor (SET) using a variety of nanostructuring approaches. However, there is also considerable worldwide activity on magnetic devices using giant magnetoresistance

(GMR) of nanostructures with architectures of various modulation dimensionalities. In fact, it is the nanostructuring with various modulation dimensionalities that has created an expanding range of different functionalities that can be engineered into these GMR devices.

There is also exciting carbon nanotube research being actively pursued in areas of high-field-emission displays and several other nanoscale electronic devices. This is an area still very early in its development, since nanotubes and their derivatives are a relatively recent discovery, but one with tremendous potential. While there is little technological impact already present in the nanoscale device area other than GMR read heads, several potential areas of significant impact do appear on the horizon. These include terabit memory and microprocessing; single molecule DNA sizing and sequencing; biomedical sensors; low-noise, low-threshold lasers; and nanotubes for high brightness displays. Nevertheless, a major challenge looms in the efficient manipulation of these nanoscale building blocks and their eventual commercial scaleup, if any of this is really going to affect society as we know it.

One shining example that indicates probable success in overcoming such obstacles in the future is the ability to now translate SET devices made by individual atomic manipulation into arrays of similarly functional devices created by the biological self-assembly of large molecular arrays. Such cross-disciplinary transfers of nanostructuring ideas and capabilities can be expected to increasingly impact the future successful implementation of nanostructure science and technology. In the area of consolidated materials, reviewed by Carl Koch, we have known for about a decade that the bulk behavior of materials can be dramatically altered when constituted of, or consolidated from, nanoscale building blocks. This can significantly and favorably affect the mechanical properties, magnetic properties, and optical properties of a range of engineering materials.

We already know that the hardness and strength of nanophase metals can be greatly increased by nanostructuring, for example. On the other hand, the ductility and superplastic forming capabilities of nanophase ceramics have now become possible generically, leading to new processing routes that will be more cost-effective than present methods. Nanoparticle fillers in metal, ceramic, or polymer matrices can yield a very wide range of nanocomposites with unique properties. This is an area that in some cases is just beginning to be researched seriously, but it could have huge technological impact in the future. Nanostructuring can also uniquely create both soft and hard magnetic materials with greatly improved performances. These materials are already having technological impact in the areas of low-loss magnets, high-hardness and tough cutting tools, and nanocomposite cements.

Potential technological applications with high commercial impact can be expected in the areas of superplastic forming of ceramics, ultrahigh-strength and tough structural materials, magnetic refrigerants, a wide range of nanoparticle-filled polymer nanocomposites based on elastomers, thermoplastics and thermosets, and ductile cements. Lynn Jelinski describes nanoparticles, nanostructured materials, and nanodevices from the point of view of biological applications and biological analogies. Current research directed toward biological synthesis and assembly is highlighted as it pertains to the building blocks of nanotechnology, and examples are presented of state-of-the-art research on the biological aspects of dispersions and coatings, high surface area materials, and functional nanostructures. A primary finding is that although biological applications of nanostructure science and technology may not be as well developed currently as non-biological ones, they nevertheless present a very promising research and development frontier that is likely to have tremendous future impact. Funding and research programs in nanotechnology around the world are reviewed by Mike Roco.

It is noteworthy that these funding levels have been increasing very rapidly in recent years as the number of researchers worldwide who are excited about this field have multiplied and funding agencies have responded accordingly. The various ways in which nanostructure science and technology research is funded in the countries the panel surveyed had often appeared quite different from a distance, but are actually quite similar to one another at closer view. Some countries, most notably Japan, have tended to primarily fund their nanostructure research through large national programs, with a rather monolithic appearance from afar, centered at national laboratories or at major national universities. On the other hand, with some exceptions, most of the nanostructure research funding in the United States and Europe tends to be based upon competition among individual research groups for smaller amounts of support.

In both types of nanostructure funding schemes, however, it seems that the individual researchers actually dominate how the work proceeds. In most cases, any significant interactions among researchers occur through normal personal and professional contacts; large-scale institutionalized cooperative research efforts in this field have often not been particularly effective. A particularly impressive national funding effort in nanostructure science and technology occurs in France under the auspices of the Centre National de la Récherche Scientifique (CNRS). There, an extensive multidisciplinary network of laboratories in universities, industries, and national laboratories, funded partly by the CNRS and partly by industry, appear to interact successfully.

The current levels of activity of the major regions assessed in this WTEC study (Europe, Japan, and the United States), for the broad areas of synthesis and assembly, biological approaches and applications, dispersions and coatings, high surface area materials, nanodevices, and consolidated materials. These comparisons are, of course, integrals over rather large areas of a huge field and therefore possess all of the inevitable faults of such an integration. At best, they represent only a snapshot of the present. Nevertheless, the panel drew the following general conclusions. In the synthesis and assembly area, the United States appears to be somewhat ahead, with Europe and then Japan following. In the area of biological approaches and applications, the United States and Europe appear to be on a par, with Japan following. In nanoscale dispersions and coatings, the United States and Europe are again at a similar level, with Japan following. In the area of high surface area materials, the United States is clearly ahead of Europe, which is followed by Japan. On the other hand, in the nanodevices area, Japan seems to be leading quite strongly, with Europe and the United States following. Finally, in the area of consolidated nanomaterials, Japan appears to be a clear leader, with the United States and Europe following. Nanostructure science and technology is clearly a very broad and interdisciplinary area of research and development activity worldwide. It has been growing explosively in the past few years, since the realization that creating new materials and devices from nanoscale building blocks could access new and improved properties and functionalities. While many aspects of the field existed well before nanostructure science and technology became a definable entity during the past decade, it has really only become a coherent field of endeavor through the confluence of three crucial technological streams:

1. new and improved control of the size and manipulation of nanoscale building blocks
2. new and improved characterization (e.g., spatial resolution, chemical sensitivity) of materials at the nanoscale
3. new and improved understanding of the relationships between nanostructure and properties and how these can be engineered.

These developments have allowed for an accelerating rate of information transfer across disciplinary boundaries, with the realization that nanostructure scientists can and should borrow insights and techniques across disciplines, and for an increased access to common enabling tools and technologies. We are now at the threshold of a revolution in the ways in which materials and products are created. How this revolution will develop, and how great will be the opportunities that nanostructuring can yield in the future, will depend upon the ways in which a number of challenges are met. Among the challenges facing nanostructure scientists and engineers in order for rapid progress to continue in this field are the necessary advances that must be made in several enabling technologies. We need to increase the capabilities in material characterization, be it in visualization or analytical chemistry, at ever finer size scales. We also need to be able to manipulate matter at finer and finer size scales, and we must eventually use computational approaches in directing this. Experiment simply cannot do it alone; theory and modeling are essential. Fortunately, this is an area in which the sizes of the building blocks and their assemblies are small enough that it is possible, with the ever increasing capabilities of computational sciences, to start doing very serious controlled modeling experiments to guide researchers in the nanostructuring of matter. Hence, multiscale modeling, across atomic, mesoscopic, and macroscopic length scales, of nanostructuring and the resulting hierarchical structures and material properties is an absolute necessity as we attempt in the coming decades to utilize the tremendous potential of nanostructure science and technology. Another challenge is to fully understand the critical roles that surfaces and interfaces play in nanomaterials, owing to the very high specific surface areas of nanoparticles and the large areas of interfaces in the assembled nanophase forms. We need to know in detail not only the structures of these interfaces, but also their local chemistries and the effects of segregation and interaction between the nanoscale building blocks and their surroundings. We also need to learn more about the control parameters of nanostructure size and size distribution, composition, and assembly. For some applications of these building blocks, there are very stringent conditions on these parameters; in other applications considerably less so. We must therefore understand the relationships between the limits of this stringency and the desired material or device properties if efficient utilization of nanostructuring is to be achieved.

Since nanostructures are often inherently unstable owing to their small constituent sizes and high chemical activity, a further challenge is to increase the thermal, chemical, and structural stability of these materials and the devices made therefrom, in the various temperatures and chemistries of the environments in which the nanostructures are asked to function. A nanostructure that is only a nanostructure at the beginning of a process is not of much use unless the process is over in a very short time or unless the process itself is the actual nanostructure advantage. So, stability is a real concern in many applications. Researchers must determine whether natural stability or metastability is sufficient or if we must additionally stabilize against the changes that we cannot afford. Fortunately, it appears that many nanostructures possess either a deeply metastable structure or they can be readily stabilized or passivated using rather traditional strategies. Reproducibility and scalability of nanoparticle synthesis and consolidation processes in nanostructuring are paramount for successful utilization of nanostructure research and development. What is accomplished in the laboratory must eventually benefit the society that pays the bills for the research, or the field will simply die. Also, significant enhancements in statistically driven process controls are required if we are to be able to effectively commercialize and utilize the nanostructuring of matter. New thinking is needed, not only about the materials, not only about the processing and assembly of these materials, but also about the manufacture of products from these materials and the economic impact of dealing with effluents. Given the commercial promise of net-shape forming of nanoscale ceramics, for example, the viability of such nanostructure production and utilization depends upon the total integrated costs of precursors or raw materials, synthesis of the building blocks, manufacturing of parts from those building blocks, and finally, disposition of the effluents. Higher than normal up-front costs for the nanoparticles or building blocks may be affordable if the processing steps save more than that. It is the total integrated costs, along with societal needs, that will determine commercial viability Education is also of tremendous importance to the future of the field of nanostructure science and technology. The creation of a new breed of researchers working across traditional disciplines and thinking “outside the box” is an absolute necessity for the field of nanostructure science and technology to truly reach fruition and to impact society with full force. The education of this new breed of researchers, who will either themselves work across disciplines or know how to work with others across disciplinary lines in the interfaces between disciplines, is necessary to make this happen in the future. People will need to start thinking in truly unconventional ways, if we are to take full advantage of this excitingly new and revolutionary field.

It appears that nanostructure science and technology at present resembles only the tip of a pyramid that has recently been uncovered from the sands of ignorance. As the new and expanding research community of nanostructure scholars worldwide digs away at these sands and uncovers more and more of the exciting field of nanostructure science and technology, we will eventually learn how truly important the field will have become and how great its impact will be on society. From our present vantage point, this future looks very exciting.

Nanoscience: an Interdisciplinary Field of Science and Technology

The Kuhnian notion of 'paradigm' is commonplace nowadays. Dosi first introduced that notion in technology studies. He assumed that 'normal' technological change consists of incremental, relatively small improvements that follow bigger, revolutionary (and therefore 'scarce') technological breakthroughs which ultimately result in new technological paradigms. According to Dosi (1982, p. 152) a technological paradigm "embodies strong prescription on the directions of technical change to pursue and those to neglect". Dosi (1988) defined a technological paradigm as a model and pattern of solution of selected technological problems, based on highly selected principles from natural sciences, jointly with specific rules aimed at acquiring new knowledge [...] A technological paradigm is both an exemplar—an artifact that is to be developed and improved—and a set of heuristics.

Since Dosi, the notion of 'technological paradigm' has been used by so many researchers that even this concept has become a commonplace. Substantial qualitative and theoretical work is available on the emergence of a new technological paradigm. Debackere and Rappa (1994) have suggested that technological paradigms typically emerge in two phases: bootlegging and bandwagon.

During the bootlegging period, which may last for a long time, a small number of researchers dedicate themselves to furthering the field. Their peers may not share their enthusiasm. Frequently, researchers from such an emerging community have to face severe criticism. Typically, they have difficulties in securing adequate funding, hence, the term 'bootlegging'. Typically, a few isolated individuals start working on similar problems with roughly similar ideas (Debackere & Rappa 1994, pp. 27-28). Researchers who are dedicated to a new and unorthodox field of inquiry often face a difficult dilemma. On the one hand, before receiving resources, they need more proof that their work will yield results. On the other hand, without resources, they are unable to precisely do that. 'Bootlegging' enables fledging research to proceed without the full knowledge and scrutiny of managers and other researchers, up to a point at which the promise of the idea is clear. During this phase then, the community will be highly concentrated among a small number of organizations, and the yearly increase in number of researchers is fairly moderate. As the number of individuals working on the same problem area increases, a communication network emerges with ties that are much stronger than the ties binding the individuals to the organizations they formally belong to. During this 2nd, so-called bandwagon phase of the community life cycle, a very rapid increase occurs in the number of researchers working in the community, with this taking place over a relative short period of time. As the community grows, a new paradigm comes into being, indicated by the higher-level network of the (sub-)discipline as competing with the older paradigm. The community tries to organize congresses and found journals, so as to be able to steer the selection process. The R&D community is typically distributed across organizations, sectors, and countries. If the work of a new community seems interesting from a commercial point of view, some scientists may be recruited by enterprises, while some who already work within industry are allowed to devote their efforts openly to the new field. Finally, some scientists may decide to become entrepreneurs themselves.

In terms familiar to the field of futures studies, one can compare the new paradigm in the bootlegging stage with a weak signal that only few take seriously. In the bandwagon stage it develops towards a strong signal that has to be taken into account.

This paper presents an overview of our theoretical work regarding leitbilds. After introducing the basic concepts we apply our heuristics to nanotechnology. Drawing on a number of technical reports on developments in nanoscience and technology we try to characterize the leitbild system of nanotechnology. We discuss the potential use of patent and publication based data to generate topics within the aforementioned leitbild systems. The paper concludes with a suggested model as to how one can evaluate the epistemic utility of a leitbild.

Technology Generalizations and Leitbilds

Technology generalizations are different types of perceived similarities between the already existing technological innovation and a potentially new technology. Similarities concern both the techniques applied in the innovation and the targets that are achieved based on the innovation (Kuusi & Meyer 2002). Two techniques are similar in the sense that they could replace each other in the achievement of (defined) targets. Another form of generalization is based on the realized techniques of the innovation that are used for new 'similar' applications.

In terms familiar to futures studies, one can compare a new paradigm in the bootlegging stage with a weak signal that only few people take seriously. In the bandwagon stage, it develops towards a strong signal that must be taken into account.

A concept that illustrates the guiding function of an emerging technological paradigm is the 'leitbild'. '*Leitbild*' is a German word. Its most general meaning is *ein Bild, das leitet*, a guiding image. According to Marz and Dierkes (1994), a leitbild has two functions, guidance and image. The guidance function consists of three subfunctions: (1) creating a shared overall goal, or 'collective projection'; (2) orientation toward one long-term overall goal, or 'synchronous preadaptation'; (3) working in the same direction, or 'functional equivalency.' The image function consists of three subfunctions: (1) cognitive activator; (2) providing a focal point, or 'individual activator'; and (3) 'interpersonal stabilizer'. Like a common vision, a leitbild creates a shared overall goal, offers orientation toward one long-term overall goal, and provides a basis for different professions and disciplines to work in the same direction. Leitbild refers not only to a common vision of actors; it also relates to the concept of autopoesis (from Greek, self-organization) and functions as an interpersonal stabilizer. With an efficient leitbild, no center is needed that urges or controls individuals to perform certain functions.

Inspired by Marz and Dierkes (1994), we characterize the general rules of an emerging paradigm as a system of leitbilds. An emerging technological paradigm is typically a system of many competing leitbilds. In the bandwagon (or paradigmatic) stage, one leitbild often begins to dominate. Leitbilds are used in visions, but it is important to distinguish between a 'leitbild' and a 'vision'. Followers of a leitbild form a kind of 'intellectual community', but as long as their visions differ, they usually do not establish a real R&D community. The intellectual community of a leitbild typically integrates several R&D communities and their members.

We use the notion of technological leitbild systems (Kuusi & Meyer 2002) to explore inter-relations and connections between seemingly separate areas, because a leitbild system can establish links through similarities or analogies. A leitbild system is a system of guiding images that create a shared overall goal, offer orientation toward one long-term overall goal, and provide a basis for different professions and disciplines to work into the same direction. Thus, a leitbild system defines the development path of a technological paradigm.

Bijker (1993) has introduced the notion of a 'technological frame' that combines the cognitive and the social sphere, including exemplary artifacts, cultural values, goals, scientific theories, and tacit knowledge. A frame is not fixed, but built up and sustained by the process of stabilizing artifacts, and is internal to the set of interactions within a relevant social group. However, actors can be members of more than one frame/social group with different degrees of inclusion in any frame. Above all, a technological frame provides "the goals, the thoughts and the tools for action", whilst at the same time limiting the freedom to act. In this way interactions create a structure that, in turn, constrains further interactions (Bijker 1993, Martin 1998).

Bijker's concept of a 'technological frame' is quite close to our understanding of leitbilds. It is an important step toward notions of technological trajectories, which are more closely related to concepts of technological determinism. However, the notion of 'technological frames' does not give technology the prominent role it deserves. Here, our leitbild concept steps in. Our concept appreciates both the importance of social factors that influence the exploration of technological options and the technological determinants that confine the relevant cognitive processes to certain research, development, and design spaces.

Types of Technological Generalizations in a Technological Paradigm

In this section we discuss how for an emerging technological paradigm, future applications can be anticipated. Kuusi suggests that a technological paradigm is a "shared generalization language" capable of producing important generalizations (Kuusi 1999). These generalizations are based on a cluster of linked technologies. The language of a promising technological paradigm can be viewed as a cluster consisting of realized and promising targets and realized and promising techniques. Realized targets are existing artifacts – or, more precisely, their properties or functions – while realized techniques are production processes and design methods. The similarity between techniques is based on the perceptions and interpretations of experts in the corresponding field, whereas the similarities between targets are based on perceptions and interpretations of the users of the artifacts. The underlying idea of the generalization concept is that existing techniques and targets serve as a platform for a process generating technological options in a multitude of ways. Generalizations are always based on perceived similarities. Emerging paradigms provide similarities based on both realized targets and realized techniques. On the other hand, a technological paradigm is the result of this type of generalization process, its successes and failures. Realized targets, which have been achieved with realized techniques ('successful exemplars'), and unsuccessful exemplars are 'concepts' of the generalization language.

Six different types of generalization. Realized techniques can be generalized so as to predict promising techniques (arrow 1), if both techniques are considered scientifically similar. From the point of view of the paradigm, there are no fundamental technical problems in Type 1 generalizations. It simply requires some effort. For example, once you have realized that a certain virus can be used to transfer a gene to a bacterium, it is reasonable to believe that you might also use another (similar) virus for that purpose. Another form of generalization is based on already realized techniques that bear a potential beyond their current range of application. Techniques can be used to create new artifacts that are (from the point of view of the paradigm) similar (arrow 2). Like Type 1, this generalization is based on scientific similarity, but only partly. For example, once you have realized that you can transfer a gene to a certain bacterium with a virus, it is reasonable to believe that you might transfer the gene in a similar way to another bacterium. But is the gene transfer to the second bacterium as acceptable to your customer as the first transfer? The targets (or the transfers) in both cases might be very similar from a technical point of view but very different from the point of view of your customer. Your customer might consider that the second transfer is irrelevant or even unethical. It is important to realize that technological paradigms as 'generalization languages' are also based on customer values. Actually, we assume in our model that similarities between targets are based only on the interpretations of customers.

Once you have realized a target or made a new artifact using a certain technique, you might start thinking about new ways to produce the artifact or new techniques to improve it (arrow 3). This is a new line for technological generalizations, or for enriching the 'paradigmatic language'. You might eventually include in your paradigm new techniques that have technically very little to do with your original techniques. Consider fusion energy. The original technical idea of the fusion bomb has very little in common with the recent techniques based on the use of huge magnets.

Generalizations of Types 1, 2, and 3 are relatively well grounded. It is possible, however, in the language of a paradigm to make generalizations that are far less grounded. Instead of strong scientific similarities, they are based on possible social developments or on weak scientific similarities (weak scientific or technical signals). One can anticipate techniques that would become promising if somebody first realizes certain targets (arrow 4). For example, if you are able to set up a permanent colony of people on the moon, new efficient ways to produce solar energy on the moon might become possible. Or you might anticipate new targets to be achieved if you could realize a technique that is supported only by weak technical signals (arrow 5). For example, if you can produce energy cheaply, you might provide an abundant supply of fresh water from salt water.

There is still one arrow in our picture left, arrow 6. It means that a person or an organization whose target B has been achieved considers that it is also possible to achieve similar target B'. How successful is this type of generalizations? Frequently such generalizations are irrational and have often resulted in questionable processes. Why are Type 6 generalizations frequently unsuccessful? The important point is that in our model – as well as in reality – the similarity between B and B' is based only on the interpretation of users of the realized artifact. In all other generalizations, similarity interpretations are either made only by technical experts or by users and technical experts together.

We illustrate our point with an example. Energy users realized in the early 1950s that it is possible to make commercial energy from atomic fission by using similar techniques of the fission bomb. Based on this generalization, many users made a Type 6 technology generalization. They considered that in similar way one could proceed from atomic fusion bombs to commercial fusion energy and provided a considerable amount of funding for the development of the commercial fusion power. Though opportunistic technical experts have used the funding for the development of commercial fusion energy, they were surely aware already in the early 1950s of huge technical difficulties of that project. In order to produce commercial fusion energy, you have to keep the fuel for a relatively long period at extremely high temperature and under equally high pressure. That is not needed in the production of energy from atomic fission. If all the money that has been used for the development of commercial fusion energy would be have been used, *e.g.*, on solar power, the energy situation of humankind might be much better.

Application of the Model to Nanotechnology

Now, we apply our approach introduced in the previous sections to the field of nanotechnology. The term 'nanotechnology' was first coined by Norio Taniguchi in the 1970s in Japan where it is associated with top-down miniaturization "which can be regarded as the latest stage in mechanical engineering, which has pursued ever-tighter precision of manufacture and tolerances throughout its history" (Budworth 1996, p. 13). In the 1980s, Drexler began to use the term nanotechnology to denote his vision of molecular manufacturing (Drexler *et al.* 1991, p. 294). The main difference between leitbild and vision is that 'vision' is an actor related concept in the framework of visionary management. Persons or organizations might have visions that give them the ability to plan or set policy in a far-sighted way. Leitbild is not related to any specific actor. It is a principle that can be selected as a part of a vision. According to Grupp (1993, p. 65), "nanotechnology will have a key position in the technological development of the 1990s and in the first decades of the 21st century". He described the field as an enabling technology that "makes possible engineering at the level of atoms and molecules" and continues: This new basic technology can stimulate future innovation processes and new generations of technologies. It is based on the interaction of information technology, polymer research, optics, biochemistry and medicine and micromechanics.

Grupp's characterization of nanotechnology indicates the early-stage character of the field, but also shows the potential it holds. His description further underlines the interdisciplinary and cross-boundary nature of the area, which provides a substantial challenge to what is perceived as necessary collaboration between sectors and disciplines. Considerable efforts from various sides have been undertaken to forecast the development of this novel field of science and technology. For instance, the German Mini-Delphi study chose nanotechnology as an explicit category.

Nano-resolution analytical methods as depicted in topics 20 and 22 can be viewed as generalizations of Type 1 – from already realized techniques to other promising techniques. The aim here is to further improve existing tools, typically in an incremental fashion, by adding new functions to analysis tools. In our example, realized techniques, such as atomic force microscopes (AFM's) or scanning-tunneling microscopes (STM's), are further generalized into promising tools that are not yet developed but conceivable from the already existing technological platforms. Further, very incremental developments of scanning force microscopes can be expected to improve the reaction and synthesis methods or chemical analysis. Along with further technical development of scanning-probe methods, researchers are discovering new phenomena in the fields of physics, chemistry, and biology. At the same time these microscopy techniques are increasingly used as a 'tool' rather than a 'probe'. The idea is to modify surfaces and tailor their structures on the nano-scale, down to the manipulation of individual atoms (Frenken, 1998, pp. 289-299). Ultimately they might facilitate large-scale manipulation at the nanometer level. However, this transcends the possibility of Type 1 generalizations.

Nanomaterials are an area that is characterized by Type 2 generalization, the transition from realized techniques to promising targets. Together with a better scientific understanding of the subject matter, a variety of already realized techniques allow developing rather specific ideas of improved materials. By taking advantage of nanoscale characteristics of structures and substances, one may create new materials with enhanced properties, such as polymers, composites, or other materials (topics 16 & 17). Rather than direct control of individual atoms, bulk operations suffice to exploit these nanoscale properties.

Another example of bulk-processing nanomaterials are colloidal dispersions (Philipse 1998, pp. 171-8). Colloid science deals with the physics and chemistry of finely dispersed particles with at least one dimension in the submicron range, including nanoparticles that are frequently considered smaller than 100 nm. Colloid science has a long tradition involving nanoparticles such that not all that is nano is necessarily new. In this sense, colloids encompass gold colloids, colloidal silica, and aluminum oxide powders. Due to their small dimensions, colloids exhibit Brownian motion. Owing to their large surface area, the interaction between colloidal particles in the liquid phase is determined by surface forces, such as Van der Waals attractions, and repulsions due to the particle charge. The balance between these forces critically depends on the details of the particle surface and the liquid composition. Colloids easily aggregate to form large aggregates, networks, or gels. While there are already techniques to control these aggregation processes to some extent, our understanding remains limited. Yet we know enough of the existing techniques and about potential ways to improve them to envisage also improved properties of materials and, ultimately, products, such as milk, cosmetics like toothpaste or sunscreen, or ink, which are nothing but suspensions of colloids or dispersions. Computer simulation and statistical mechanics are tools that are used to further understand colloidal systems.

Thin-film techniques are an example of Type 4 generalization from promising targets to promising techniques. Realized techniques already permit sufficiently exact operations at the nanometer level to suggest the idea of future products that would require even more exact and precise tools. This generalization requires a preceding Type 2 generalization. Thin-film technologies are a considerably well-developed field. The ultra-fine production of thin films is necessary for the subsequent characterization.

Designing ultra-thin layers is associated with a number of aims, such as atomically exact delineations of layers, quantized potential distribution, defined pore distribution in layers, ultra-thin separation and protection layers, and improved layer function by way of multilayer structuring. These targets are in turn motivated by and related to many technical applications, including information storage layers, films with quantum effects, optical layers, multilayer piles for semiconductor laser and X-ray optical compounds, displays, sensor layers, tribologic films, biocompatible films, photovoltaic films, membrane films, and chemically active surfaces (Bachmann 1998), which are the starting point for Type 4 generalizations toward new, improved techniques.

Two topics in our Delphi example correspond to this type of generalization (topics 14 & 18). Here efforts appear to be directed at characterizing these structures. Topic 14, for instance, suggests that the control of monomolecular layers will allow developing organic hybrid composite materials. The aim of controlling monomolecular layers, while not yet possible, is based on the progress made with existing tools and techniques that allow speculating about the properties of new products or processes, which in turn leads to the next step towards improved instruments.

The topics in the area of biomimetics (19, B) are examples of Type 3 generalization from realized targets to promising techniques. The idea is to simulate nature in order to develop materials with novel properties by way of self-organization. The biomimetic approach can be used as a path to obtaining novel materials, using self-assembly techniques to make organic templates on which inorganic structures are then deposited (Budworth 1996, p. 7).

While basic principles of self-organization are known, we still need to integrate various techniques to achieve the target of controlled self-assembly. Although one can create structures by way of self-organization in a biomimetic process, our technological means are still incomplete to fully utilize the potential this leitbild offers. Being aware of the general feasibility – thanks to already realized artifacts – we can make reasonable assumptions about the requirements of the techniques necessary to pursue this path of development further.

Topics 15, 21, and 23 in the Delphi study describe a leitbild that focuses on the direct control of atoms in order to rearrange them to form new structures that could result in novel materials. This leitbild follows a Type 5 generalization, from promising techniques to promising targets. Building on Type 1 generalization, it is first based on the availability of promising techniques from which promising targets are then projected. As pointed out in leitbild I, we can reasonably expect current STM and AFM technologies to be further developed into more complex tools that, beyond measurement and observation, can efficiently manipulate structures at the nanometer scale. From such promising technique one can make the Type 5 generalization step to improved and novel artifacts.

The difference between the materials approach, leitbild II, and leitbild V is the different control of processes, bulk reactions versus atomic control. Atomic control is also strongly related to the idea of atoms being effectively used as carrier of certain functions, such as data storage, *etc*.

The Leitbild System of Nanotechnology

All the different approaches we call leitbilds belong to one greater whole that eventually will develop into a technological system. As long as the exact shape of that technological system is unclear, we speak of a leitbild system instead. One element of this leitbild system might even substitute and outdate another leitbild. For instance, what we identified as leitbild V could replace II one day. Even though both approaches refer to nanostructures, they are essentially different. While II uses bulk methods, V aims at direct atomic control. A leitbild and, even more so, a leitbild system is coined by the integration of a number of communities. Even though leibild II is a field that is relatively close to realization, it still critically relies on the integration of knowledge from a variety of disciplines and of expertise from a number of industrial sectors. For instance, even for monitoring and controlling activities at the bulk level, it is necessary to use nano-resolution instruments. The borderlines between science and engineering disciplines become blurred, and disciplinary fields tend to fuse as in the field of materials science and engineering. This is even more apparent in the area of biomimetics, which tries to simulate natural principles to build up structures. At the nanometer level, the boundaries between disciplines tend to disappear. This is why we can refer to nanotechnology as a leitbild *system* that integrates different approaches, each of which being autonomous enough to bear its own identity, but also depending to a greater or lesser extent on results from the other fields.

Promoty Technological Generalizations Emerging Leitbilds

People committed to different leitbilds considerably differ in their evaluations of the future prospects of generic technologies. How can we make different evaluations/ interpretations more explicit? Kuusi (1999) has suggested that that we can handle the difference by measuring the *epistemic utility*. The idea is that for an actor it is more reasonable to start a realization process of a certain option, if the epistemic utility of that option increases. In the bootlegging stage of a leitbild, there are only few actors who believe in the reasonability of the underlying generalizations. Most experts think that the generalizations will not be realized at all or that it takes too long before it is reasonable to start the realization process. If the leitbild has proceeded to the bandwagon stage, a majority of actors believe in rather quick realization of the generalizations. The epistemic utility of the topic has increased dramatically for average actors. Any new successful generalization of the emerging technology presented during the process between the bootlegging stage and the bandwagon stage has some impact on this growth of the epistemic utility. In this study, we will not discuss how to measure epistemic utility (see Kuusi 1999). It is sufficient to mention four aspects of the epistemic utility of a technological generalization. The epistemic utility is related, first, to the anticipated impacts of the generalization; second, to the value (positive or negative relevance) given by relevant stakeholders to different impacts; and, third, to the techniques available for the realization of the generalization. Typically a champion of a technological generalization has in the bootlegging stage much more positive evaluation concerning these aspects than mainstream actors. The important fourth aspect is the evaluated validity of the three anticipated aspects.

National technology foresight Delphi studies have had 'proxy' measures for the variables of the four aspects of the epistemic utility. The degree of the importance of each topic has been measured by the Delphi panelists' evaluations (Cuhls & Kuwahara 1994, NISTEP 2001), which refer to our first two aspects: the impacts and their relevance. The evaluation scales exclude topics being evaluated feasible but undesirable, which implies the questionable assumption that the realization of topics is always desirable, though more or less important. In the latest Japanese Technology Foresight study, the impacts are also discussed with expected effects and potential problems of technology generalizations (NISTEP 2001). Evaluated effects are socio-economic development, resolution of global problems, people's needs, and expansion of intellectual resources; potential problems are adverse effect on the natural environment, on safety, and on morals/culture/society. Proxy measures for feasibility are the anticipated cost constraint as well as technical, funding, human resources, and R&D system constraints on technological generalizations (Cuhls & Kuwahara 1994). Two proxy measures for the validity of an evaluation are the degree of certainty of an expert concerning the realization time of a topic and the self-evaluation of the expertise (Loveridge *et al.* 1995, NISTEP 2001). Evaluations of the epistemic utility of technological generalizations also provide a heuristics for the decision making of a company. Let us suppose that a company includes only one champion of a technology generalization based on an emerging paradigm who considers starting the realization project a reasonable choice, which means that only for him or her the epistemic utility sufficiently high. The managers of that corporation could base their decision in favor of the project on two reasonable necessary conditions: (1) the champion is a reasonable person; and (2) the champion is ready to take an economic risk with this project. If these two conditions are met, a reasonable choice for the firm would be to start a new venture with the champion. This strategy has been empirically found *e.g.* by Lovio (1993) in the Finnish electronic industry in the 1980s. Another reasonable policy is to allow the champion to continue the bootlegging as long as the epistemic utility is growing both for the champion and other key persons in the company. This means that the champion has to produce new arguments (*e.g.* realized minor generalizations) which step by step convince new protagonists.

With respect to forthcoming research activities, we approached the question as to how to generate candidates for leitbilds from data on the current research and technology. In the early 1990s, patent data was used in mid-term oriented Foresight activities

(Grupp 1993). With respect to nanotechnology, more recent work was carried out by Meyer *et al.* (2002). Using bibliometric techniques with patent and publication data allows filtering and identifying core concepts that emerge in a specific area. Mapping an area over time can illustrate when new concepts have emerged and may allow speculation on what new technological steps can be expected. Using elements of our leitbilds, experts may be able to identify clusters of techniques that would allow addressing some promising targets or conversely could speculate on how nanoscale techniques currently under development could be extended in their area of application. However, keyword maps are typically limited to a set of the top 60 or so concepts that occur most frequently and are therefore by default fairly general in nature. Instead of focusing on the top 60 concepts, we plan to investigate a subset of nanotechnology areas (nanobiotechnology, nano-structured materials and surface characterization) to generate a set of more specific concepts from which experts could generate topics suitable for a Delphi study. We assume to find candidates for different leitbilds by applying cluster analysis to second order concepts in the patent applications (*e.g.* ranks 100-200).Another application of bibliometric techniques would be the identification of potential experts, based on mostly cited or linked documents in the leitbild system candidates. Interviews with these experts may allow further analysis of their key technology generalizations and leitbilds.

Biology, Chemistry, Physics, Engineering and Computer Science

Most progress in nanoscale science and technology results from research involving various combinations of Biology, Chemistry, Physics, Engineering and Computer Science. Therefore, a basic understanding of each of these subjects is a valuable asset for a person engaged in interdisciplinary nanoscience. A diverse background will give a nanoscientist the ability to communicate with colleagues and find the appropriate methods for a particular project. While each of these subjects has a plethora of available information from a macroscale perspective, there is plenty of room for their development with respect to the nanoscale. With such a background established, a strong focus on the nanoscale intersection between two or more of these fields is a promising method for achieving success in the field of nanotechnology.

Biology provides proof that the concept behind nanotechnology is realistic. The dis-assembly and re-assembly of biological systems into their individual components such as *DNA*, *protein* and *phospholipids* is common practice for the molecular biologist and has resulted in numerous advancements in *biotechnology*.

Chemistry is a well established method for dealing with atoms and molecules in very precise statistical sense. *Nanoscale chemistry* involves the study of single molecules, molecular assembly lines, nanoscale reaction vessels and more.

Physics has produced tools that can be used to directly interact with the nanoscale, namely *Scanning Probe Microscopy (SPM)*. SPM technology was predicted in the 1950s by *Feynman* and its discovery in the 1980s could be considered the spark that started.

Engineering on the nanoscale involves the use of atomically precise *components* to design and build nanoscale devices. Simulation and fabrication of *nanomachines*, *quantum computers* and *molecular electronics* may soon become standard practice in the *engineering community*.

Computer Science is a field that may be the first to develop real nanotechnology. Rapidly *shrinking computer chips* and *novel architechtures* require new chip fabrication methods and new software. *Molecular simulation* is an essential tool for nanotechnology.

Perhaps the most promising approach to nanotech education is to explore various combinations of the above fields. For instance, chemical engineering is a well established and profitable combination. Merging biology with computer science results in the important subject of bioinformatics. For my own undergraduate education, I chose a biochemistry major with a minor in mathematics. Meanwhile, our understanding of the nanoscale is rapidly expanding into numerous disciplines as scanning probe methods are developed by physicists.

Nanoscience

The Department of Defense (DoD) has selected six research activities in its Basic Research Plan (BRP) that may offer the highest payoff for meeting future military needs. Strategic Research Objectives (SROs) have been defined for each of these six activities. Nanoscience is one of these SROs. Nanoscience is pervasive across the other five SROs. Advancements in this area cannot help but have direct or indirect benefits to the others. In fact, it might be said that nanoscience is at the frontier of modern science and technology.

Simply stated, nanoscience is concerned with how atoms go together to create molecules of interest to human existence. Information will be presented generally on the six Strategic Research Objectives and specifically on nanoscience. Emphasis will be placed on efforts in nanoscience and applications to guidance and control. Strategic Research Objectives A brief description of each of the six Strategic Research Objectives contained in the Department of Defense Basic Research Plan is given as follows: *Nanoscience*. Achieve dramatic, innovative enhancements in the properties and performance of structures, materials, and devices that have controllable features on the nanometer scale (one billionth of a meter).

The payoffs include: ultra-small computers with terabit nonvolatile random access memories and teraflop speeds; image information processors; low-power personal communication devices; synthesis of new materials for advanced electronic, magnetic, and optical sensors; and, significant life-cycle cost reductions in many systems through failure remediation.

Biomimetrics. Enable the development of novel synthetic materials, processes, and sensors through advanced understanding and exploitation of design principles found in natural biological systems. The payoffs include advanced adhesives, infrared signature visualization, sensing of only a few molecules of a chemical, armor protection, production of designer vaccines and pharmaceuticals, novel gene therapy, and new detectors for environmental monitoring.

Smart Structures. Demonstrate advance capabilities for modeling, predicting, controlling, and optimizing the dynamic response of complex, multi-element, deformable structures used in land, sea, and aerospace vehicles and systems. The payoffs include: more accurate rapid fire weapon systems; sensor arrays for assessing the operational environment; control of conformal antennas, phased arrays, and broadband spiral antenna systems; shock, noise, vibration, and stability isolation, augmentation, and control of submarines, surface ships, helicopters, and aircraft; and reliable actuator materials and lightweight structural materials capable of physical and virtual shape changes.

Mobile Wireless Communication. Provide fundamental advances enabling the rapid and secure transmission of large quantities of multimedia speech, data, graphics, and video information from point to point, broadcast and multicast over distributed networks for heterogeneous C3I systems. Payoffs include: increased radio carrier frequencies that will enable wider bandwidth; lower power, antenna beam steering; improved throughput, survivability, and security of complex mobile communication networks; and, integrated cable, satellite, and mobile wireless communications.

Intelligent Systems. Enable the development of advanced systems able to sense, analyze, learn, adapt, and function effectively in changing or hostile environments until completing assigned missions and functions. The payoffs include a complex series of advancements across the board in science that will provide seamless information flow among people, sources, and automated systems such as robots, software, and computing modules.

Compact Power Sources. Achieve significant improvements in the power and energy density, operating temperature, reliability, and safety of compact battery, fuel cell, and other compact power sources through the identification and exploitation of new concepts. The payoffs mainly emphasize saving millions of dollars a year for power sources for all the electronic gadgets on the battlefield. Nanotechnology Nanoscience may be what the Department of Defense calls this SRO, but nanotechnology is where the action is. As stated earlier, one nanometer is a billionth of a meter (10-9 meter), abbreviated as nm. However, nanotechnology as a technology is concerned with dimensions of a micrometer (10-6 meter), only a millionth of a meter, down to one tenth of a nanometer (10-10 meter). Nanotechnology covers four orders of magnitude from the sizes of atoms to molecules to the aggregation of molecules into materials. The structures of polymers, DNA, viruses, and other biological moieties fall within the nanotechnology spectrum. Life is first exhibited at the nanotechnology level. The beginning of nanotechnology is often attributed to a talk.

Another milestone in nanotechnology was the pioneering source book on nanotechnology by K. Eric Drexler, *Nanosystems: Molecular Machinery, Manufacturing, and Computation,* that was published by John Wiley & Sons in 1992. He succinctly describes the unique nature of nanotechnology with his discussion of "top-down" versus "bottom-up" approaches. A wide gap has separated the top-down path of microtechnology (starting with large, complex, and irregular structures) from the bottom-up path of chemistry (starting with small, simple, and exact structures). In microtechnology, the challenge is to make imprecise structures smaller; in chemical synthesis, the challenge is to make precise structures larger. An engineering discipline is only now forming around the latter goal. The top-down approach is that used in making sophisticated electronic devices.

Discrete transistors, Small Scale Integrated Circuits (SSIC), Large Scale Integrated Circuits (LSIC), and Very Large Scale Integrated Circuits (VLSIC) decrease in size from about one millimeter (10-3 meter) to onetenth of a micrometer (10-7 meter) in average feature sizes. This amounts to about an order of magnitude change for each type of device. All of these microelectronic circuits have used chemical vapor deposition (CVD), molecular beam epitaxy (MBE), or some related technique to deposit layers of different materials that are later carved using a photo-lithography mask and electronic beams or x-rays. Current state-of-the-art has reached about 350 nm. The goal is to make "point-one micron" (10-7 meter), 100 nm, feature size devices. Many consider that this may be the limit to using conventional microelectronic manufacturing techniques. This is the top-down approach to nanotechnology. The bottom-up approach is the creation of structures beginning at one-tenth nanometer and adding to feature sizes. The goal of nanotechnology is to manipulate atoms and build molecules and subsequently to combine molecules into supramolecules, devices, structures, and micro-machines.

This manipulation is primarily a function of molecular positional control. The precision of positional control is dependent upon thermal, quantum, and mechanical vibrations. Drexler makes an interesting comment in his book that is relevant to nanoscience: "In most sciences, theoretical work is scarce relative to experimental work (the abundant output of theoretical physics is an outstanding exception)." The science of nanotechnology involves atomic forces, molecular bonding, quantum phenomena, and the shadow world of Van Der Waals forces and mysterious force potentials. Except for quantum theories, the description of nanotechnology is mostly empirical. Theoretical physics is actively looking at the "Casimir effect" which may explain some of the phenomena taking place across the spectrum of nanotechnology. Progress in nanoscience and nanotechnology was reported at the *Fifth Foresight Conference on Molecular Nanotechnology* on 6-8 November 1997 in Palo Alto, California. Over 300 attendees heard 95 presentations and a panel discussion. Several foreign countries attended and presented papers: Australia, Austria, Belarus, Denmark, Finland, Germany, Italy, Japan, The Netherlands, Poland, Russia, Siberia, Sweden, and Ukraine.

The primary objectives of the papers could be summarized as follows: link theory with practice; build a self-assembler for self-replication; and, develop programmable position control. The keynote speaker was Dr. Richard Smalley, who won the 1996 Nobel Prize in Chemistry for the discovery of complex carbon molecules called fullerenes (an Army Research Office project). Fullerene structures are also known as "bucky balls." The carbon structure was named after R. Buckminster Fuller who designed geodesic dome structures. The carbon molecular form with 60 carbon atoms was named "buckminsterfullerene" because of its structural resemblance to a geodesic dome. The name "fullerene" was given to any closedcage molecule containing an even number of carbon atoms. The fullerene structure can be converted into nanometer-size cylinders called nanotubes. Nanotubes are essentially wires on a molecular scale with a very high Young's modulus in the neighborhood of a tera-Pascal. Nanotubes could also carry a very high current. Smalley is the head of the Center for Nanoscale Science and Technology at Rice University.

The various forms of nanotubes were the subject of a number of presentations. A different structure of carbon atoms is exhibited by diamond. The primary interest in theoretical studies of molecular nanotechnology has been on tetrahedral carbon frameworks called diamondoid structures. Tetrahedral frameworks may also be made out of silicates, substituted silicates, and related molecules, although this is a relatively new area in molecular nanotechnology. Fullerene technology is the primary path to molecular nanotechnology. A number of devices are described and illustrated at *http: / / www.mitre.org/ research/ nanotech*. Experiments on nanometer-range materials and phenomena are very dependent upon tools for positioning and measurement. The imaging and manipulating of atoms, molecules, and fragments of large molecules have been made possible with the development of proximal probes such as: the electric force microscope (EFM), atomic force microscope (ATM), scanning probe microscope (SPM), and scanning tunneling microscope (STM).

Several papers described computer models and supercomputer simulations of designs of various types of rotating gears made out of molecules of chiefly hydrogen, carbon, silicon, nitrogen, phosphorous, oxygen, and sulfur. One source for information on the design of these gears is the Institute of Molecular Manufacturing. Some excellent simulations were described by NASA, who have a major nanotechnology effort. Potential applications include aerospace vehicles, computers, cooling, power, and system architectures. NASA was a cosponsor of the Foresight Conference. Another group of papers was concerned with nanometersize biological structures. Carbon has a fundamental role in the natural nanosystems of biochemistry. The material of choice is DNA which is 2 nm across and 10 nm long. It is possible to design many arrangements of DNA fragments. One application was a DNA chip for diagnostic purposes. See *http: // mmtsb.scripps.edu*. The Foresight Institute was founded twelve years ago, specifically with the objective of advancing nanotechnology. The Sixth Foresight Conference on Molecular Nanotechnology was held in Santa Clara, California on 13-15 November 1998.

General access to the Foresight Institute may be achieved at *http: / /www.foresight.org*. Many options on nanotechnology are offered. *Nanotechnology Magazine*, which has been closely associated with the Foresight Institute, can be reached at *http: / / www.nanozine.com*.

Guidance And Control Applications Guidance and control components in precision guided munitions should be a major beneficiary from advancements in nanoscience and nanotechnology. The potential exists for miniaturized sensors and surveillance packages, efficient night vision devices, new antenna designs, terabit memory density, ultra fast logic, faster optoelectronics, zero threshold lasers, innovative circuit design concepts, and micro-electromechanical devices. Smart munitions can become brilliant munitions with the help of nanotechnology. An excellent review of the progress that might be expected in the top-down approach is contained in a 1997 report, *Energy-Efficient Technologies for the Dismounted Soldier,* National Academy Press, Box 285, 2101 Constitution Avenue, Washington, DC 20055. This study looks at improving the performance of man-portable weapon and support systems from two perspectives: improve the available power supplies, or design electronics that require less power. Both perspectives are relevant to guidance and control, but the second approach contains an excellent assessment of the trends in electronic devices. This is particularly important because electronics represent about 60 percent of the cost of a precision guided munition. Projections of the performance of electronic devices indicate possibly two or three orders of magnitude advancement before encountering fundamental limits in the nanometer range.

The present objective is to achieve low power gigascale integration (GSI), a billion transistors per chip by 2010. The specific target is one-billion gate, 100 nm, 1.0 GHz GSI technology. Beyond that is the terascale. The earliest payoff for guidance and control applications will come from top-down efforts to reduce the feature size of microelectronics. Scaling down feature sizes provides the most significant indicator of improved performance and increased productivity in guidance and control components. The feature sizes of chips will be about 250 nm by the year 2000 and 100 to 125 nm by 2010. After that, further progress appears to come face-to-face with progress that has to be made in the bottom-up approach. Theoretical limits have been defined for the top-down approach.

Five different levels of limits have been defined to date: fundamental limits; material limits; device limits; circuit limits; and system limits. Fundamental limits are represented by random thermal motion of carriers in solids, effects of the Heisenberg uncertainty principle on switching energy, and the fact that pulses cannot exceed the speed of light. Material limits are set by semiconductor materials and interconnect materials. Device limits are set by the switching energy and the delays inherent in present device designs. Circuit limits are primarily controlled by interconnects. System limits are imposed by chip architecture, switching energy of semiconductor technology, energy storage and removal, operating cycle time, and by the size of an individual chip. Evaluations of such theoretical limits can provide hints of the directions that future science and technology should take. It would appear that the projected performance enhancements and cost reductions contained in the Defense Technology Area Plan for guidance and control technology are more feasible based upon the expectations of nanotechnology.

A case study should be performed on the impact of specific nanotechnology accomplishments on a selected precision guided munition. Simultaneously pursuing the bottom-up and the top-down approaches is like building a tunnel through a mountain from both sides until the builders meet approximately in the middle. It is obvious that the bottom-up builders are starting in more difficult terrain. Progress in science and technology will depend upon the support of both approaches.

Natural Law & Science

Science and technology intertwine. Engineers use knowledge produced by scientists; scientists use tools produced by engineers. Scientists and engineers both work with mathematical descriptions of natural laws and test ideas with experiments. But science and technology differ radically in their basis, methods, and aims. Understanding these differences is crucial to sound foresight. Though both fields consist of evolving *meme* systems, they evolve under different pressures. Consider the roots of scientific knowledge.

Through most of history, people had little understanding of evolution. This left philosophers thinking that sensory evidence, through reason, must somehow imprint on the mind all human knowledge-including knowledge of natural law. But in 1737, the Scottish philosopher *David Hume* presented them with a nasty puzzle: he showed that observations cannot logically prove a general rule, that the Sun shining day after day proves nothing, logically, about its shining tomorrow. And indeed, someday the Sun will fail, disproving any such logic. Hume's problem appeared to destroy the idea of rational knowledge, greatly upsetting rational philosophers (including himself). They thrashed and sweated, and irrationalism gained ground. In 1945, philosopher *Bertrand Russell observed* that "the growth of unreason throughout the nineteenth century and what has passed of the twentieth is a natural sequel to Hume's destruction of empiricism." Hume's problem-meme had undercut the very idea of rational knowledge, at least as people had imagined it.

In recent decades, Karl Popper (perhaps the scientists' favorite philosopher of science), Thomas Kuhn, and others have recognized science as an evolutionary process. They see it not as a mechanical process by which observations somehow generate conclusions, but as a battle where ideas compete for acceptance.

All ideas, as memes, compete for acceptance, but the meme system of science is special: it has a tradition of deliberate idea *mutation*, and a unique immune system for controlling the mutants. The results of evolution vary with the selective pressures applied, whether among test tube *RNA* molecules, insects, ideas, or machines. Hardware evolved for refrigeration differs from hardware evolved for transportation, hence refrigerators make very poor cars. In general, replicators evolved for A differ from those evolved for B. Memes are no exception.

Broadly speaking, ideas can evolve *to seem true or they can evolve to be true* (by seeming true to people who check ideas carefully). Anthropologists and historians have described what happens when ideas evolve to *seem* true among people lacking the methods of science; the results (the evil-spirit theory of disease, the lights-on-a-dome theory of stars, and so forth) were fairly consistent worldwide. Psychologists probing people's naive misconceptions about how objects fall have found beliefs like those that evolved into formal "scientific" systems during the Middle Ages, before the work of Galileo and Newton.

Galileo and Newton used experiments and observations to test ideas about objects and motion, beginning an era of dramatic scientific progress: Newton evolved a theory that survived every test then available. Their method of deliberate testing killed off ideas that strayed too far from the truth, including ideas that had evolved to appeal to the naive human mind.

This trend has continued. Further variation and testing have forced the further evolution of scientific ideas, yielding some as bizarre-seeming as the varying time and curved space of relativity, or the probabilistic particle wave functions of quantum mechanics. Even biology has discarded the special life-force expected by early biologists, revealing instead elaborate systems of invisibly small molecular machines. Ideas evolved to *be* true (or close to the truth) have again and again turned out to *seem* false—or incomprehensible. The true and the true-seeming have turned out to be as different as cars and refrigerators.

Ideas in the physical sciences have evolved under several basic selection rules. First, scientists ignore ideas that lack testable consequences; they thus keep their heads from being clogged by useless parasites. Second, scientists seek replacements for ideas that have failed tests. Finally, scientists seek ideas that make the widest possible range of exact predictions, The law of gravity, for example, describes how stones fall, planets orbit, and galaxies swirl, and makes exact predictions that leave it wide open to disproof. Its breadth and precision likewise give it broad usefulness, helping engineers both to design bridges and to plan spaceflights. The scientific community provides an environment where such memes spread, forced by competition and testing to evolve toward power and accuracy. Agreement on the importance of testing theories holds the scientific community together through fierce controversies over the theories themselves. Inexact, limited evidence can never *prove* an exact, general theory (as Hume showed), but it can *disprove* some theories and so help scientists choose among them. Like other evolutionary processes, science creates something positive (a growing store of useful theories) through a double negative (*dis*proof of *in*correct theories). The central role of negative evidence accounts for some of the mental upset caused by science: as an engine of disproof, it can uproot cherished beliefs, leaving psychological voids that it need not refill. In practical terms, of course, much scientific knowledge is as solid as a rock dropped on your toe. We know Earth circles the Sun (though our senses suggest otherwise) because the theory fits endless observations, and because we know why our senses are fooled. We have more than a mere theory that atoms exist: we have bonded them to form molecules, tickled light from them, seen them under microscopes (barely), and smashed them to pieces. We have more than a mere theory of evolution: we have observed mutations, observed selection, and observed evolution in the laboratory. We have found the traces of past evolution in our planet's rocks, and have observed evolution shaping our tools, our minds, and the ideas in our minds—including the idea of evolution itself. The process of science has hammered out a unified explanation of many facts, including how people and science themselves came to be. When science finishes disproving theories, *the survivors often huddle so close together* that the gap between them makes no practical difference. After all, a practical difference between two surviving theories could be tested and used to disprove one of them. The differences among modern theories of gravity, for instance, are far too subtle to trouble engineers who are planning flights through the gravity fields of space. In fact, engineers plan spaceflights using Newton's disproved theory because it is simpler than Einstein's, and is accurate enough. Einstein's theory of gravity has survived all tests so far, yet there is no absolute proof for it and there never will be. His theory makes exact predictions about everything everywhere (at least about gravitational matters), but scientists can only make approximate measurements of some things somewhere. And, *as Karl Popper points out*, one can always invent a theory so similar to another that existing evidence cannot tell them apart. Though media debates highlight the shaky, disputed borders of knowledge, the power of science to build agreement remains clear. Where else has agreement on so much grown so steadily and so internationally? Surely not in politics, religion, or art. Indeed, the chief rival of science is a relative: engineering, which also evolves through proposals and rigorous testing.

As IBM Director of Research Ralph E. Gomory says, "The evolution of technology development is often confused with science in the public mind." This confusion muddles our efforts at foresight.

Though engineers often tread uncertain ground, they are not doomed to do so, as scientists are. They can escape the inherent risks of proposing precise, universal scientific theories. Engineers need only show that under *particular* conditions *particular* objects will perform *well enough*. A designer need know neither the exact stress in a suspension bridge cable nor the exact stress that will break it; the cable will support the bridge so long as the first remains below the second, whatever they may be.

Though measurements cannot prove precise equality, they *can* prove inequality. Engineering results can thus be solid in a way that precise scientific theories cannot. Engineering results can even survive disproof of the scientific theories supporting them, when the new theory gives similar results. The case for assemblers, for example, will survive any possible refinements in our theory of quantum mechanics and molecular bonds.

Predicting the content of new scientific knowledge is logically impossible because it makes no sense to claim to *know* already the facts you will *learn* in the future. Predicting the details of future technology, on the other hand, is merely difficult. Science aims at knowing, but engineering aims at doing; this lets engineers speak of future achievements without paradox. They can evolve their hardware in the world of mind and computation, before cutting metal or even filling in all the details of a design.

Scientists commonly recognize this difference between scientific foresight and technological foresight: they readily make *technological* predictions about science. Scientists could and did predict the quality of Voyager's pictures of Saturn's rings, for example, though not their surprising content. Indeed, they predicted the pictures' quality while the cameras were as yet mere ideas and drawings. Their calculations used well-tested principles of optics, involving no new science.

Because science aims to understand how everything works, scientific training can be a great aid in understanding specific pieces of hardware. Still, it does not automatically bring engineering expertise; designing an airliner requires much more than a knowledge of the sciences of metallurgy and aerodynamics.

Scientists are encouraged by their colleagues and their training to focus on ideas that can be tested with available apparatus. The resulting short-term focus often serves science well: it keeps scientists from wandering off into foggy worlds of untested fantasy, and swift testing makes for an efficient mental immune system. Regrettably, though, this cultural bias toward short-term testing may make scientists less interested in long-term advances in technology. The impossibility of genuine foresight regarding science leads many scientists to regard all statements about future developments as "speculative"—a term that makes perfect sense when applied to the future of science, but little sense when applied to well-grounded projections in technology. But most engineers share similar leanings toward the short term. They too are encouraged by their training, colleagues, and employers to focus on just one kind of problem: the design of systems that can be made with present technology or with technology just around the corner. Even long-term engineering projects like the space shuttle must have a technology cutoff date after which no new developments can become part of the basic design of the system. In brief, scientists refuse to predict future scientific knowledge, and seldom discuss future engineering developments. Engineers do project future developments, but seldom discuss any not based on present abilities. Yet this leaves a crucial gap: what of engineering developments firmly based on *present science* but awaiting *future abilities*? This gap leaves a fruitful area for study. Imagine a line of development which involves using existing tools to build new tools, then using *those* tools to build novel hardware (perhaps including yet another generation of tools). Each set of tools may rest on established principles, yet the whole development sequence may take many years, as each step brings a host of specific problems to iron out. Scientists planning their next experiment and engineers designing their next device may well ignore all but the first step. Still, the end result may be foreseeable, lying well within the bounds of the possible shown by established science. Recent history illustrates this pattern. Few engineers considered building space stations before rockets reached orbit, but the principles were clear enough, and space systems engineering is now a thriving field. Similarly, few mathematicians and engineers studied the possibilities of computation until computers were built, though many did afterward. So it is not too surprising that few scientists and engineers have yet examined the future of *nanotechnology*, however important it may become.

Efforts to project engineering developments have a long history, and past examples illustrate present possibilities. For example, how did Leonardo da Vinci succeed in foreseeing so much, and why did he sometimes fail? Leonardo lived five hundred years ago, his life spanning the discovery of the New World. He made projections in the form of drawings and inventions; each design may be seen as a projection that something much like it could be made to work. He succeeded as a mechanical engineer: he designed workable devices (some were not to be built for centuries) for excavating, metalworking, transmitting power, and other purposes. He failed as an aircraft engineer: we now know that his flying machines could never be made to work as described. His successes at machine design are easy to understand. If parts can be made accurately enough, of a hard enough, strong enough material, then the design of slow-moving machines with levers, pulleys, and rolling bearings becomes a matter of geometry and leverage. Leonardo understood these quite well. Some of his "predictions" were long-range, but only because many years passed before people learned to make parts precise enough, hard enough, and strong enough to build (for instance) good ball bearings—their use came some three hundred years after Leonardo proposed them. Similarly, gears with superior, cycloidal teeth went unmade for almost two centuries after Leonardo drew them, and one of his chain-drive designs went unbuilt for almost three centuries. His failures with aircraft are also easy to understand. Because Leonardo's age lacked a science of aerodynamics, he could neither calculate the forces on wings nor know the requirements for aircraft power and control.

Can people in our time hope to make projections regarding molecular machines as accurate as those Leonardo da Vinci made regarding metal machines? Can we avoid errors like those in his plans for flying machines? Leonardo's example suggests that we can. It may help to remember that Leonardo himself probably lacked confidence in his aircraft, and that his errors nonetheless held a germ of truth. He was right to believe that flying machines of some sort were possible-indeed, he could be certain of it because they already existed. Birds, bats, and bees proved the possibility of flight. Further, though there were no working examples of his ball bearings, gears, and chain drives, he could have confidence in their principles. Able minds had already built a broad foundation of knowledge about geometry and the laws of leverage. The required strength and accuracy of the parts may have caused him doubt, but not *their interplay of function and motion*. Leonardo could propose machines requiring better parts than any then known, and still have a measure of confidence in his designs.

Proposed molecular technologies likewise rest on a broad foundation of knowledge, not only of geometry and leverage, but of chemical bonding, statistical mechanics, and physics in general. This time, though, the problems of material properties and fabrication accuracy do not arise in any separate way. The properties of atoms and bonds *are* the material properties, and atoms come prefabricated and perfectly standardized. Thus we now seem better prepared for foresight than were people in Leonardo's time: we know more about molecules and controlled bonding than they knew about steel and precision machining. In addition, we can point to nanomachines that already exist in the *cell* as Leonardo could point to the machines (birds) already flying in the sky. Projecting how second-generation nanomachines can be built by protein machines is surely easier than it was to project how precise steel machines would be built starting with the cruder machines of Leonardo's time. Learning to use crude machines to make more precise machines was bound to take time, and the methods were far from obvious. Molecular machines, in contrast, will be built from identical prefabricated atomic parts which need only be assembled. Making precise machines with crooked machines must have been harder to imagine then than molecular assembly is now. And besides, we know that molecular assembly happens all the time in nature. Again, we have firmer grounds for confidence than Leonardo did.

In Leonardo's time, people had scant knowledge of electricity and magnetism, and knew nothing of molecules and quantum mechanics. Accordingly, electric lights, radios, and computers would have baffled them. Today, however, the basic laws most important to engineering—those describing normal matter—seem well understood. As with surviving theories of gravity, the scientific engine of disproof has forced surviving theories of matter into close agreement. Such knowledge is recent. Before this century people did not understand why solids were solid or why the Sun shone. Scientists did not understand the laws that governed matter in the ordinary world of molecules, people, planets, and stars. This is why our century has sprouted transistors and hydrogen bombs, and why *molecular technology* draws near. This knowledge brings new hopes and dangers, but at least it gives us the means to see ahead and to prepare. When the basic laws of a technology are known, future possibilities can be foreseen (though with gaps, or Leonardo would have foreseen mechanical computers). Even when the basic laws are poorly known, as were the principles of aerodynamics in Leonardo's time, nature can demonstrate possibilities. Finally, when both science and nature point to a possibility, these lessons suggest that we take it to heart and plan accordingly.

X-rays and Neutrons: Essential Tools for Nanoscience Research

Virtually all of the grand challenges in nanotechnology have as overriding themes the need to determine the structure over a wide range of length scales and to understand how the combination of the structure and dynamics leads to their unique properties. Both x-rays and neutrons cover all the length scales of interest in nanoscience, from the atomic structure of individual building blocks to the configuration of assembled, functional structures, making them essential tools for the elucidation of these challenges.

The workshop is one of a series of workshops in support of the National Nanotechnology Initiative (NNI) Strategic Plan prepared by the Nanoscale Science and Engineering and Technology Subcommittee (NSTC), recently updated in December 2004 as part of the 21st Century Nanotechnology Research and Development Act (Public Law 108-153).

As called for in the Strategic Plan, the workshop will provide a guiding strategy to policy makers for the development of essential scattering tools for R&D in nanotechnology. The workshop report will be developed by internationally renowned participants who will identify frontier problems in their fields from experimental and theoretical perspectives by addressing the essential questions:

- Which of the outstanding problems in nanoscale synthesis, structure, dynamics and properties can be addressed using X-ray and neutron techniques such as scattering, imaging and spectroscopy, and how can these techniques help illuminate the important and urgent issues at the nanoscale?
- How might the current resource base in X-ray and neutron scattering techniques be augmented and used in solving outstanding problems in nanoscale science?

A scholarship program will also provide the opportunity for selected early career researchers to participate fully in the workshop and the development of the ensuing report.

The workshop will bring together established experts and young scientists from the x-ray, neutron, and nanoscience/nanotechnology communities to discuss future research directions and opportunities. The goal of the workshop is to identify and promote the innovations in scattering techniques and related instrumentation required to tackle the exciting challenges of nanotechnology research now and for the future. The focus of this workshop will be to identify frontier problems in nanoscience such as understanding interfacial structures, nano-systems, confinement, and self-assembly of hard materials, soft materials, and biomaterials which can be elucidated through the use of x-ray and neutron scattering techniques.

After these research opportunities are identified, the workshop report will outline a roadmap for the prioritized development of these critical resources. Specifically, the workshop report will address the following goals:

- Identify the challenges in characterization at the nanoscale for the next 5-10 years that may be addressed using x-ray or neutron scattering techniques.
- Identify the instrumentation and techniques that must be developed to meet these challenges and allow research in nanoscience to advance.
- Identify both short and long term R&D in areas such as beam optics, detectors and in-situ characterization that will be required to support this vision.

The workshop report will be an effective tool for funding agencies, principal investigators, academia and industry to coordinate their long term investment strategies.

- U.S. Department of Energy, Office of Basic Energy Sciences
- National Institute of Standards and Technology, NIST Center for Neutron Research
- National Science Foundation, Division of Materials Sciences

Workshop Chairs:

- Ian Anderson (Oak Ridge National Laboratory, Spallation Neutron Source)
- Linda Horton (Oak Ridge National Laboratory, Center for Nanophase Materials Sciences)
- Eric Isaacs (Argonne National Laboratory, Center for Nanoscale Materials, and University of Chicago)
- Mark Ratner (Northwestern University)

Workshop Organizers:

- Ian Anderson (Oak Ridge National Laboratory, Spallation Neutron Source)
- Kristin Bennett (U.S. Department of Energy, Office of Basic Energy Sciences)
- Al Ekkebus (Oak Ridge National Laboratory, Spallation Neutron Source)
- Pat Gallagher (National Institute of Standards and Technology)

- Phillip Lippel (National Nanotechnology Coordination Office)
- Linda Horton (Oak Ridge National Laboratory, Center for Nanophase Materials Sciences)
- Eric Isaacs (Argonne National Laboratory, Center for Nanoscale Materials)
- Helen Kerch (U.S. Department of Energy, Office of Basic Energy Sciences)
- Celia Merzbacher (Office of Science and Technology Policy)
- Mark Ratner (Northwestern University)
- Guebre Tessema (National Science Foundation, Division of Materials Research)

The Nanosurf® Mobile S

The Mobile S atomic force microscope is a whole new way to look at atomic force microscopy. Mobility. Ease of use. Sleek design. Modern and well-integrated software. Unsurpassed support and reliability. And a price much lower than you'd pay for a comparable AFM system. These are features that cannot be matched by any other AFM. This revolution in design increases your productivity, decreases your downtime, and gives you more time to focus on your real application issues. Being an expert in AFM is no longer required to get results. The mobile S features the most commonly used modes of AFM operation. Dynamic Force or "tapping mode", contact mode, force modulation, phase contrast imaging, surface spreading resistance, and force spectroscopy are among the supported modes. Additionally, user inputs and customized scripting is supported with all modes. Integrated Dual-View cameras, elimination of laser and detector alignment, and intuitive software sets the Mobile S apart from all other atomic force microscope systems.

Nanosceince Instruments

Nanoscience Instruments is your one source for a wide variety of AFM systems, probes, accessories, and related nanoscience tools. Our products include the easy-to-use Nanosurf® easyScan SPMs, to the advanced WITec AlphaSNOM systems, as well as a wide variety of AFM probes, AFM add-on accessories, and post processing software for virtually every SPM system.

Nanotechnology: Science Fiction

In 1997, Francis Collins, the spokesperson for the US Human Genome Project (HGP), claimed that, "the project's Ethical, Legal and Social Implications (ELSI) program [was] unique among technology programs in its mandate to consider and deal with these issues alongside the development of the technology" (cited in McCain 2003, p. 112). Recent assessments of its impact have been far from celebratory (*e.g.* Evans 2002, Huijer 2003, McCain 2003). Indeed, it is claimed that the ELSI program insulated the HGP from criticism rather than facilitating negotiations between scientists and non-scientists (Huijer 2003, p. 488).

Against this background, when Mihail Roco, a key promoter of nanoscience and technology (NST) in the U.S. and director of the National Nanotechnology Initiative (NNI), claims that societal implications have been an integral component of the NNI from the start and argues that the National Science Foundation (NSF) "has made support for social, ethical and economic research studies a priority" (Roco 2003a, p. 185), it is reasonable to wonder to what extent this represents a genuine invitation to the agora or a façade that merely disguises science's traditional agoraphobia. As Nik Brown has recently argued, "The 'post-normal science' thesis [...] which sees science increasingly dependent on wider political and public aspirations should, it appears, be received with caution" (Brown 2003, p. 18). However, notwithstanding these reservations, it would be wrong to dismiss the opportunities created by the current social and political exigencies requiring technoscience to explore the ethical and social implications of its activities. Even if it is just a façade, it represents a surface that at the very least can be tagged with critical graffiti.

Having said this, it would be equally problematic to think that the deliberative space in which discussions of the social and ethical implications of nanotechnology are unfolding, or will unfold, is an empty one. At the moment, this space is being structured by a form of extrapolation that draws on narrative elements from the science fiction (SF) genre. In this paper, I argue that there are important limitations associated with trying to understand the ethical and social implications of NST within this discursive space, and that an understanding of these limitations must precede or should be taken into account in a more reflexive debate on NST. First, I consider the implications of arguing that SF is not an external but an internal aspect of NST discourse. Following this, I show that the central metaphor in NST discourse – nanotechnoscientists as master builders – provides a semantic link to SF narrative elements. This link allows NST authors to extrapolate by drawing on SF world-building techniques. I, then, provide a detailed analysis of this process by examining the key role that the SF literary device of the *novum* plays in two NST texts. A number of scholars have already drawn attention to the important function that the SF literary device of the *novum* plays in NST discourse, especially Milburn (2002) and Marshall (2004) and to a lesser extent Miksanek (2001) and Landon (2004). What my discussion adds to these is a more detailed textual analysis of the functioning of the *novum* and an exploration of the implications of these discursive strategies for debates on the ethical and social implications of NST.

The first text is Drexler's *Engines of Creation*, the second an edited book on the convergence of nano, bio, information technology and cognitive science (NBIC), *Converging Technologies for Improving Human Performance*. It is edited by Mihail C. Roco and Williams Sims Bainbridge, both active promoters of NST. Although many in the NST community might argue that Drexler's vision is both dated and outside the mainstream, the editors of and contributors to the second text are very much part of the NST mainstream. In bringing these two texts together, I show that there are more similarities than would be initially expected, not necessarily in terms of their substantive claims but in terms of the formal narrative structures through which their claims are engendered. This is followed by a discussion of the limitations associated with the framing of ethical and social implications by current NST discourse. In the conclusion, I consider some further implications of the way NST discourse mobilizes the future.

Technoscientists in the NST field frequently draw on SF in order to construct a binary opposition that is deployed to police the boundary (Gieryn 1999) between science and non-science. A well-known instance is the debate initiated by Gary Stix (1996), a staff writer for *Scientific American*, who wrote a highly critical piece on Eric Drexler's agenda for nanotechnology. Amongst other things, as Milburn notes, Stix compares "Drexler's writing to the scientific romances of Jules Verne and H.G. Wells, suggesting that 'real nanotechnology' is not to be found in these science fiction stories" (Milburn 2002, p. 265). However, as Fogelberg and Glimell argue in their analysis of the debate, one of the key issues at stake is the meaning of scientific practice (Fogelberg & Glimell 2003, pp. 10-12). Drexler's more speculative extrapolative approach, based on theoretical computational modeling, is seen to be at odds with the experimentally based work that is taken to be the hallmark of good science.

My purpose in examining the relationship between SF and NST is not to explore how SF is invoked to criticize NST for its 'unscientific excesses' or to address the mediating role that SF, as an object external to science, might play between the scientific community and the public at large in popular culture. Instead, I argue that narrative elements from the SF genre are not external to but contribute to the constitution of NST discourse itself. By drawing attention to the shifting and permeable border between science and SF in NST, it is not my intention to either put in question the scientific credentials of nanotechnoscientists by insinuating that they are not doing 'real science' or, more generally, to undermine the credibility of science due to its reliance on narrative techniques found in fiction. As Donna Haraway argues, "Not only is no language, including mathematics, ever free of troping; not only is facticity always saturated by metaphoricity; but also, any sustained account of the world is dense with storytelling. 'Reality' is not compromised by the pervasiveness of narrative; one gives up nothing except the illusion of epistemological transcendence, by attending closely to stories." [Haraway 1997, p. 64] Science is not possible despite narrative but precisely because of it. However, not all narratives are the same; they draw on different naratological devices. Discourses that extrapolate technoscientific developments into the future, through SF narrative elements, contain assumptions about, amongst other things, the nature of being, the dynamics of historical change, the aspirations of citizens, and the relationship between society, culture and technoscience. With this in mind, I will now discuss more specifically how SF narrative elements are incorporated into NST.

Definitions of NST are highly contested (Fogelberg & Glimell 2003, pp. 5-26). This is due to its status as an 'emergent science', that is to say a science whose truth claims remain to be settled by scientific or public consensus (Hamilton 2003, p. 268). What is more, given the heterogeneous and interdisciplinary nature of the NST field, it is not likely that definitional closure will be achieved soon. Definitional problems also arise because NST, as we shall see below, is radically future oriented; thus, it is also defined by how its potential is refracted towards competing futures (Wood *et al.* 2003, p. 3). However, a minimum definition would draw attention to the significance of its length scale:

> One nanometer (one billionth of a meter) is a magical point on the dimensional scale. Nanostructures are at the confluence of the smallest of human-made devices and the largest molecules of living things. Nanoscale science and engineering here refer to the fundamental understanding and resulting technological advances arising from the exploitation of new physical, chemical and biological properties of systems that are intermediate in size, between isolated atoms and molecules and bulk materials, where the transitional properties between the two limits can be controlled. [Roco cited in Ratner & Ratner 2003, p. 7] The 'newness' of the nanoscale refers to the difference between the macroscopic and nanoscopic properties of materials. To take Ratner and Ratner's (2003) example, although a metric ton, a kilogram, and gram of gold all have the same physical properties, the same is not true when one scales down to the nano length. Gold's color, melting point, and chemical properties are different at the nano length scale as a result of the nature of atomic interactions and the fact that these are not averaged out as they are in bulk material. In other words, "Nano gold doesn't act like bulk gold" (Ratner & Ratner 2003, p. 2). Thus as Roco and Bainbridge argue, "The nanoscale is not just another step toward miniaturization, but a qualitatively new scale. The new behavior is dominated by quantum mechanics, material confinement in small structures, large interfaces, and other unique properties" (2001, pp. 4-5). When these unique properties are combined with the prevalent and dominant metaphor that reigns in NST discourse – nanostructures as the *building blocks* of matter and the nanotechnoscientist as the *master builder* – we can begin to appreciate the radical transformative powers that NST not only denotes but also connotes. For instance, the Nobel laureate for physics, Horst Stormer, suggests that when we are empowered by nanotechnology to "play with the ultimate toy box of nature – atoms and molecules [...] the possibilities to create new things appear limitless" (cited in NSTC 1999, p. 1). Indeed it is not infrequent to encounter references to NST's radical transformative powers. For instance, claims such as "By anyone's measure, nanotechnology is the next big thing. In fact, according to government R&D planners, nanotechnology is nothing short of the next Industrial Revolution" (Schulz 2000, p. 41) are rather common.

In fact, these claims are foundational for analyses of the ethical, legal, and social implications that have been initiated by the NST community itself. One might even argue that the fact that the social implications have been central to the NNI represents not so much a belief in the legitimacy of submitting NST to social and ethical analysis as much as the conviction that NST is like no other technoscientific practice in its ability to impact and transform both the social and natural world. As Wood *et al.* argue in their review of the emerging field, "Nanotechnology is being heralded as a new technological revolution, one so profound that it will touch all aspects of human society" (2003, p. 1). Yet, while it is certainly the case that there have been important developments that make possible the manipulation of matter with precision at the nanoscale, as commentators have also noted, NST is only here as a trace of a future yet to be produced (Fogelberg & Glimell 2003, Milburn 2002, Wood et al. 2003). Even NST's most energetic promoters have to admit that "nanotechnology is still in its infancy, because only rudimentary nanostructures can be created with some control" (Roco & Bainbridge 2001, p. 1). There is a rather significant gap between what can be achieved with NST today and what is imagined that will be achievable in the future; predictions of revolutionary transformations seem premature. This gap, of course, is not specific to NST and can be found in other fields.

It is typically sutured rhetorically through hype that not only mobilizes meaning but also social, political, and economic resources by promising breathtaking advances, miracle cures, and virtually unimaginable wealth. However, in the case of NST the hype is different. For instance, although biotechnology hype promises wealth, global food abundance, and intimations of immortality through genetic therapy and enhancement, NST's hype promises more! By drawing on the metaphor of the nanotechnoscientist as the master builder and NST as the toolbox that makes possible the manipulation of the fundamental stuff that makes up the world, NST claims nothing less than to be able to rebuild the world.

This ultimate conceit, which feeds NST's molecular speculations, is elegantly captured by the title of the U.S. National Science and Technology Council brochure on nanotechnology: "Nanotechnology: Shaping the World Atom by Atom" (NSTC 1999). It is instructive to compare the semantic suppleness associated with the metaphors used in the HGP with those deployed in NST. The HGP promised to produce a "plan", "blueprint", "encyclopaedia", or a "program" of life (Rothman 1998, p. 25). In all instances its vision of the future was limited by the frontier between the organic and the inorganic. However, NST's 'shaping-the-world-one-atom-at-a-time' metaphor makes it possible to transcend this boundary: Nobel laureate and nanotechnoscientist Richard Smalley claims, "Nanotechnology is the builder's final frontier" (cited in NSTC 1999, p. 1).Recognizing the centrality of the world-building metaphor is important because its semantic connotations also make it possible for NST discourse to draw on narrative elements from the SF literary genre that is, in part, characterized by its ability to produce radically different future or parallel worlds. This creates the discursive conditions for what, following Landon, we can call SF thinking. SF thinking generates the rhetoric that bridges the gap between the givens of science and the goals of the imaginary marvelous, the emphasis always on 'explaining' the marvelous with rhetoric that makes it seem plausible, or at least not yet impossible. [cited in Gerlach & Hamilton 2000, p. 465] By incorporating SF thinking, NST discourse overcomes the gap between what is possible today and what might be possible in the future. This is achieved by using extrapolative narrative techniques that are well established in the genre. In other words, or in other worlds, it is able to solve the tension inherent in claiming that we are already living in a nano-era while also recognizing that the dawning of the nano-era depends on much that has yet to happen.

Although Drexler is credited with having coined the term 'nanotechnology', it is undeniable that his status in the field is problematic. As noted above, he is one of the targets of the science fiction/fact opposition deployed to locate certain NST activities outside the boundary of real science. It is frequently argued that his vision of nanotechnology remains peripheral and outside the mainstream. As evidence for this, one might point to the fact that his agenda has certainly not been explicitly endorsed by the NNI. Still, even though his book, *Engines of Creation*, may be frequently criticized, it has also introduced a generation of scientists and engineers to a nanotech 'futurescape'. Thus even a staunch critic such as Smalley, who believes that there are insurmountable objections to Drexler's proposed molecular assemblers, has conceded that Drexler "has had tremendous effect on the field through his books" (cited in Milburn 2002, p. 280). Moreover, the disagreement between Smalley and Drexler is not about the revolutionary or social transformative impact of NST (*i.e.* its capacity to rebuild the world): "Smalley acknowledges that nanotechnology, even in the more modest form of his own nanotubes of carbon, eventually 'may change the future of humankind' and that nanotechnology from chemistry on a nanometerscale 'may make even Drexler blush'." [Fogelberg & Glimell 2003, p. 19] In making these points, it is not my intention to shore up Drexler's scientific credibility or to undermine those of his opponents. Rather, it is to note that we should not allow the controversies over the viability, or not, of molecular assemblers to obscure the similarities that exist in terms of how SF narrative elements are used to negotiate the gap between current technoscientific capabilities and their future development.

In *Engines of Creation*, Drexler introduces us to a future where molecular manufacturing will be capable of making "virtually anything from common materials without labor, replacing smoking factories with systems as clean as forests. They will transform technology and the economy at their roots [...] They will indeed be engines of abundance." [Drexler 1990, p. 63]

Engines of health, or cell repair machines, will cure disease and prolong life; other engines will contribute to the launching of a new space program. All of this, and more, will be possible on the journey towards a positive-sum society that will culminate in an "open future of wealth, room and diversity, [where] groups will be free to form almost any sort of society they wish, free to fail or set a shinning example for the world" (Drexler 1990, p. 237). The only problem is that these "engines of creation" or molecular assemblers have yet to be produced. Still, Drexler's writing narrates their coming as unavoidable.

The arrival of assemblers is to follow a path already initiated by current bio and molecular technology. Protein machines will combine the cutting and pasting abilities of enzymes with the programmability of ribosomes to produce new nanoscale non-protein materials that will in turn be used to create second-generation nanomachines or *universal assemblers*. When these assemblers are combined with the astronomical computing power of nanocomputers, the knowledge of molecular and atomic architecture collated by nano-reverse-engineering-machines or *disassemblers*, and the ability to self-replicate in order to achieve economies of nanoscale, the nano-era will finally be upon us (Drexler 1990, pp. 3-20).

If we leave aside the technical argument regarding the viability of molecular assemblers, there are a number of narrative devices that make nanomachines both credible and inevitable in Drexler's text. Following SF convention, his text constructs a "sublime chronotope" in which the action unfolds (*i.e.* the romance of how molecular assemblers will rebuild the world). SF critic Istvan Csicsery-Ronay, Jr. defines the sublime chronotope as a "literary 'space-time' where fictional things work according to their own particular laws of time and space. SF works generally depict one or more special chronotopes that are wonderfully strange and ultimately vast and powerful" (Csicsery-Ronay 1996, p. 386). In Drexler's text, the chronotope has two dimensions: synchronic (*i.e.* at one point in time) and diachronic (*i.e.* across time or historical). Synchronically, Drexler's nano-chronotope invites us to see a world that has been thoroughly overhauled and reconstituted through the tropes of the atomic, the molecular, and the machinic. Diachronically, he narrates an 'engine' of historical change that links the past with a de-familiarized present. This present promises the future, as the acorn promises the oak.

In synchronic mode, the chronotope is woven through the narration of a space that is both familiar and alien. It is our world but its landscapes, scale, structures, rules, and action are all atomic. If Marx claimed that "Value [...] does not have its description branded on its forehead; it rather transforms every product of labor into a social hieroglyphic" (cited in Graham 2002, p. 236), Drexler might argue that value is inscribed as an atomic hieroglyphic. Indeed, he begins his book by claiming that "throughout history, variations in the arrangement of atoms have distinguished the cheap from the cherished, the diseased from the healthy" (Drexler 1990, p. 3), and he goes on to write, "Our ability to arrange atoms lies at the foundation of technology. We have come far in our atom arranging, from chipping flint for arrowheads to machining aluminum for spaceships" (*ibid.*).

Framed by Drexler's nano-chronotope, human interaction with nature and the development of technology is nothing more than the attempt to manipulate atoms, initially clumsily but increasingly with more precision (*i.e.* bulk versus molecular technology). The chronotope that stages the plot in *Engines of Creation* not only invites us to reconsider our relationship vis-à-vis nature, it also demands that we develop a molecular conception of our bodily selves: "The ill, the old and the injured all suffer from misarranged patterns of atoms, whether misarranged by invading viruses, passing time or swerving cars. Devices able to rearrange atoms will be able to set them right" (*ibid.*, p. 99). The figure of 'the machine' is the second key discursive element in the chronotope.

Drexler argues that there is no more incontrovertible evidence of the viability of nanomachines or molecular assemblers than the existence of protein machines or ribosomes that assemble proteins in our cells (*ibid.*, p. 6). Thus, "molecular machines in the cell demonstrate that molecular machines work" (*ibid.*).

A little later in the text, he first redefines life as a "special structure" which is governed by the "machinery of life" (*ibid.*, p. 17) and adds, "The history of life is the history of an arms race based on molecular machinery" (*ibid.*, p. 26). By combining a world where reality is reduced to myriads of atomic configurations and the bonds that hold or fail to bind them together, where technology is crude or precise atomic manipulation, health a harmonious atomic arrangement and disease an atomic cacophony, with vital molecular machinery as the basis of life, Drexler's text re-ontologizes the world. In doing so, he creates a sublime chronotope that provides an ideal habitat for his nanomachines because they straddle the two significant dimensions of this new domain, *i.e.* the atomic and the *machinic*. Grab your atomic force microscopes. We have entered the age of assembler and the book of the world is written in the language of atomic bonds.

The synchronic dimension of the chronotope is traversed by a diachronic or historical vector. It narrates how we have arrived at the stage where the assembler revolution is already contained in our present, making it inevitable. Drexler claims that it is possible to isolate the principles of change whose explanatory domain span "molecules, cells, beasts, minds, and machines [and] should endure even in an age of biotechnology, nanomachines and artificial minds" (*ibid.*, p. 21). After identifying molecular replicators – *i.e.* RNA, viral genes, human genes, *etc.* – as the chronotope's principal historical actors (*ibid.*, p. 25), he argues that through the evolutionary mechanisms of mutation and selection there is a continuity between *The Rise of the Replicators* (RNA molecules) and the rise of all other things that populate the earth: "Mutation and selection of genes has, through long ages, filled the world with grass and trees, with insects, fish and people. More recently other things have appeared and multiplied – tools, houses, aircraft, and computers. And like the lifeless RNA molecules, this hardware has evolved." [*Ibid.*, p. 30]

He further embeds the production of 'hardware' within evolutionary semantics by arguing that the principles of engineering can be understood in terms of mutation and selection: "In engineering, enlightened trial and error, not the planning of flawless intellects, has brought most advances" (*ibid.*, p. 31).

If synchronically we have seen how the atomic and machinic nano-chronotope provides an ideal space for molecular assemblers, diachronically the chronotope locates the engineer as the hero whose practice embodies the principles of change that govern the nano-chronotope. Moreover, by inserting the engineer in the context of evolutionary trans-historical forces, the molecular-assembler revolution becomes unstoppable. Thus, it is not surprising that he concludes his book by interrogating the present with questions that originate in the future: "If we succeed (and if you survive) then you may be honored with endless questions from pesky great-grandchildren: 'What was it like when you were a kid, back before the Breakthrough?' and 'What was it like growing old?' and 'What did you think when you heard the Breakthrough was coming?' and 'What did you do then?' By your answers you will tell once more the tale of how the future was won." [*Ibid.*, p. 239] This is not only a call for 'nano-engineers of the world to unite' and take their place in a world historical event that has already been determined, it also provides the key to the functioning of the chronotope.

What Drexler's text achieves, unwittingly or not, is a narrative that re-ontologizes the past, present, and future. This is achieved by rebuilding the world synchronically and diachronically around the *Breakthrough*, the arrival of the molecular assembler. The narrative process whereby a single element is used as the axis around which a future alternative world is generated is a key discursive element of SF. Though there is much debate of the status of SF as a genre, there is some consensus on the centrality of the device of the *novum*: "A *novum* is a deliberately introduced change made to the world as experienced by author and reader, but a change based on scientific or other logic; it is such a significant part of the SF that the *novum* frequently determines the subsequent narrative." [James 1994, p. 108, italics added]

The novum is a variation of the "What if..." question that is used as a world-building device by extrapolating the potential ramifications of the interruption to reality contained in the question (*e.g.* time travel, artificial intelligence, a parallel universe, molecular assembly). The assembled world derives its coherence not from the logic or validity of the *novum* itself but from the way all its dimensions have been processed by the machinery of the *novum*. This is precisely the discursive scaffolding that underpins Drexler's sublime chronotope and in turn provides the stage that projects nanotechnology into the future, or retracts the future into the present. In other words, this is how Drexler bridges the gap between what is possible now and what he envisions will be possible in the future. Without this SF discursive device, Drexler's vision of nanotechnology could not be assembled.

Converging Technologies for Improving Human Performance: Nanotechnology, Biotechnology, Information Technology and Cognitive Science is a published report that derives from a 2001 workshop sponsored by the NSF and the Department of Commerce (DOC). Since then a number of NBIC meetings have taken place. The report is of interest for a variety of reasons. First, it is the most recent and sustained effort to construct a broad vision of NST by making NBIC central to the achievement of a variety of technoscientific, social, economic, and political goals. Moreover, given the scope and transdisciplinary nature of the NNI, and the variety of agencies that it mobilized, a broad integrative vision is likely to remain a crucial element in future national NST initiatives. Second, the editors and contributors are drawn from NST's mainstream. For Mihail Roco, a key figure in the NNI, NBIC represents a continuation of the work already begun. Finally, both editors have consistently championed 'analyses of ethical and social implications'; consequently, it provides an ideal site to read the framing of these questions within NST discourse. NBIC supporters argue that the impetus for convergence is driven by "the integration and synergy of the four technologies (nano-bio-info-cogno) [that] originate from the nanoscale, where the building blocks of matter are established" (Roco & Bainbridge 2003, p. vii). The integration between bio and nano is possible because the unity of matter at the nanoscale means that the structure of both organic and inorganic materials is determined by the same fundamental principles. Consequently, it becomes possible for technology to "harness natural processes to engineer new materials, biological products, and machines from the nanoscale up to the scale of meters" (*ibid.*, p. 2). The integration and synergies between info nano and bio are diverse. On the one hand, the enhancement of computing power (*i.e.* speed and memory) is expected to derive from new nano-engineered materials as well as from novel architectures in the form of quantum and biological (DNA based) computing (Theis 2001; Ratner & Ratner 2003, pp. 130-39; Wood *et al.* 2003, pp. 19-24). On the other hand, developments in NST and biotechnology themselves depend on computer based modeling and visualization made possible by the digitalization of molecular processes (Johnson 2003, Thacker 2004, Roco 2003b).

The integration of the cognitive science component is tied to the development of the "Human Cognome Project" whose goal would be to map "the structure and function of the mind" (Bainbridge 2003, p. 97). It is argued that cognitive science would be able to explain "the mind and human behavior" by understanding their "physico-chemical-biological processes at the nanoscale" (Roco 2003c, p. 301). This would be made possible by the convergence of bio, computer, and nanotechnology (Roco & Bainbridge 2003, p. 12). In turn, the ability to enhance cognition and communication, as a result of the accrued knowledge, would make new scientific and technological discoveries possible. Ultimately, the multiple synergistic pathways in NBIC herald a new renaissance based on a comprehensive understanding of the structure and behavior of matter from the nanoscale up to the most complex systems yet discovered, the human brain. Unification of science based on unity in nature and its holistic investigation will lead to technological convergence and a more efficient social structure for reaching human goals. [*Ibid.*, p. 1]

It is interesting to note that amongst the goals reported in the volume are items that would not be out of place even in Eric Drexler's nano-chronotope. If Drexler entices us with visions of abundance, a contributor to the NBIC volume defines poverty as a technological challenge and predicts that intelligent machines will "eradicate poverty and usher in a golden age for all humankind" (Albus 2003, p. 293). Indeed, such will be the magnitude of the wealth produced that "new economic theories based on abundance may emerge to replace current theories based on scarcity" (*ibid.*, p. 292).

Engines of Creation, as noted above, presents the possibility of harnessing the design principles and mechanisms of biological molecular machines to create nanomachines capable of producing inorganic materials. In similar fashion, Roco and Bainbridge report that "fundamental knowledge about molecular-level processes essential to the growth and metabolism of living cells may be applied, through analogy, to development of new organic materials" (Roco & Bainbridge 2003, p. 11). If the previous two scenarios are conceivable in a Drexlerian world, the next one, brain-to-brain communication, is nudging towards SF, even by Drexler's standards (Drexler 1990, p. 234). The NBIC program foresees the development of *The Communicator*, a device that will enhance individual attributes and remove barriers to group communication such as [...] user's physical disabilities, language differences, geographic distance and disparity in the knowledge possessed by group members [...] Improving group interactions via brain-to-brain and brain-machine-brain interfaces will also be explored. [Albus *et al.* 2003, p. 276]

For Roco and Bainbridge, brain-to-brain communication provides a stepping stone towards a networked society capable of sustaining "a global intelligence" (Roco & Bainbridge 2003, p. 22) where "humanity would become like a single distributed and interconnected 'brain' based on new core pathways of society" (*ibid.*, p. 6). There are more examples that could be cited, but the point is not to isolate individual objects or scenarios that appear to be plucked out of SF novels, but to understand how NST discourse can circulate such inflated future currency as current technoscientific tender.

Not unlike Drexler's book, the NBIC text embeds its promissory notes in a sublime chronotope that re-ontologizes the world both synchronically and diachronically. In synchronic mode, the NBIC-chronotope is constituted as a continuous and unified space-time, which stretches from the nanoscale to the scale of meters and beyond (*ibid.*, p. 2). The coherence of this space is underwritten by the unity of matter at the nanoscale and is thus regulated by a hierarchy of causality that operates from the bottom up (*ibid.*). The fundamental properties of matter are determined by its constituent molecular dynamics. Phenomena such as memory, emotion, and thought are to be explained by reference to a hierarchy that privileges the nanoscale organization of atoms and constructs a causal explanatory pathway that links nanostructures to the structure of DNA that in turn extends the link to the interaction of neurons in the brain (*ibid.*, p. 13). The great chain of being begins at the bottom but does not end with the body or the brain. In the NBIC-chronotope, the conceptualization of the brain serves to fuse what would seem to be different phenomenological domains (*i.e.* the material and the social-cultural), allowing NBIC to expand its ontological prospecting claims. The brain is operationalized as a communicative and information processing system; social interaction and group behavior are defined by the same operators. This opens the way to re-ontologizing the brain's neural network as the cognitive 'nanostructure' of social life through which the bottom-up causal hierarchy can be replicated in the social and cultural domains. Thus, the ability to enhance these functions (*i.e.* information processing and communication) by drawing on a cognitive science leveraged by bio, nano, and information technology makes it possible to conceive of the "improvement of collective behavior and productivity" (Roco 2003b, p. 82). It is in this sense that convergence would lead to devices like the *Communicator* that would provide the basis for a "more efficient social structure for reaching human goals" (Roco & Bainbridge 2003, p. 1) and even for envisioning the "bond of humanity driven by an interconnected virtual brain of the Earth's communities searching for intellectual comprehension and conquest of nature" (Roco 2003b, p. 93).

Fuelling this vision of the ability to manage everything from the nanoscale to the interactions of humanity as a whole is an explicit and profound reductionism that arises from the conviction that "all disciplines share a common ability to work at the molecular and nano length scales using information technology and biology concepts" (Roco 2003b, p. 93). As a result, "partisans" who argue for the "independence of biology, psychology, and the social sciences [...] against 'reductionism', asserting that their fields had discovered autonomous truths that should not be reduced" are utterly self-defeating (Roco & Bainbridge 2003, p. 13). Underpinning this reductionism is an ontology in which distinct phenomenological domains lack domain-specific principles of organization, thus "a networked society of billions of human beings" is to the human being what "a human being is to a single nerve cell" (*ibid.*, p. 22). Consequently, a "collective social system may be compared to a larger form of a biological organism" (*ibid.*). The unification of the natural and social sciences would make possible the development of an explanatory regime capable of encompassing "collective behavior in physics, chemistry, biology, engineering, astronomy, and society" (Roco 2003b, p. 84). Indeed, in a chronotope where every internal ontological border is disassembled to its constituent molecular configurations, it becomes possible to conceive of a "predictive science of society and to apply corrective actions based on the convergence ideas of NBIC", and to re-ontologize culture as the product of the brain's physiology thus leading to the dual evolution of human culture and physiology (Roco & Bainbridge 2003, p. 22).

Underpinning the NBIC-chronotope are two distinct but interlinked tropes: 'communication' and 'unity'. That the ontologization of DNA as an informational code and the conceptualization of the genome as a biological computer (Thacker 2004, p. 40) makes it possible to think of molecular intervention in terms of reprogramming, is already well established. By asserting the unity of matter at the nanoscale and erasing the distinction between the organic and inorganic, NST is able to extend the informational paradigm to all matter: "Programmable matter is a technical approach to the physical world in which the distinction between information and materiality is effaced. For nanotech, the entire apparatus of nanomachines [...] is itself built out of the same molecular and atomic elements that compose the physical world. Nanotech's ultimate engineering fantasy – that of the nanocomputer or a computer hardware apparatus that is assembled from atoms – is a direct example of its will to materialize information." [Thacker 2004, p. 138]

NBIC incorporates society and culture into this informational logic by conceptualizing the brain itself as the programmable matter that underpins social behavior and interaction. Thus there is no longer an outside of the NBIC chronotope. The corollary to this unified world is the existence of a common molecular syntax: once deciphered, it will make just about anything possible. It is this universal machine language of matter that allows the conversion between bits, atoms, neurons, and genes and the seamless integration of people, technologies, societies, and humanity.

If synchronically it is the ontological unity of the social and natural world arising from a common molecular grammar that makes NBIC convergence inevitable, diachronically it is the ontologization of the history of humanity as a transhistorical quest for improvements in human performance. This is presented diagrammatically with a table which begins with the development of the cell, body, brain, *etc.* includes universities, printing, the industrial revolution, *etc.* and inexorably moves to NBIC in order to predict "societal and business reorganization" and even "evolution transcending human cell, body and brain" (Roco & Bainbridge 2003, p. 23). This historical trajectory opens up the space for a new type of historical actor, a new renaissance man or woman, the scientist engineer, capable of mastering the unified language of the world, in others words capable of punctuating the current equilibrium by completing a process that has already begun: NBIC convergence. The sense of inevitability, however, is the product of how the gap between the present and future has been overcome. Like Drexler's text, the NBIC text draws on the narrative device of the *novum* (*i.e.* NBIC convergence) that constructs a discursively coherent world that stretches from the past to the future. However, the price of coherence is that everything must be traced back to the interruption that the *novum* introduces. Thus, all of history converges towards the *novum* that in turn gives birth to the future. In the narrative developed in the NBIC text, every dimension of the world, both diachronic and synchronic, has been processed through the NBIC filter and colored by the trope of convergence and unity. Thus, the principles governing the structure of organic and inorganic matter converge, the technologies of different disciplines converge, the natural sciences converge, the natural and social sciences converge, individuals and technology converge, individuals converge into networks, societies converge, and humanity finally becomes unified. Environmental degradation, poverty, disease, cultural misunderstanding, war, *etc.* can all be solved through NBIC convergence. The NBIC world is akin to a hall of mirrors with NBIC convergence at the center: though stretched, contorted, and deformed, every reflection refers back to the principle of NBIC. It is precisely this that makes the extrapolated future credible.

The reasons why the device of the *novum* fails to generate a propitious context for the consideration of the ethical and social implications of NST are the very same reasons that explain its success and centrality as a narrative device in the literary genre of SF. In the later, its function is the construction of a coherent and plausible world that is separated from our own world in time and/or space. This is achieved by making the *novum* the crucible on which all aspects of the extrapolated world are forged. It functions through a viral logic by replicating itself in all the principal phenomenological domains of the chronotope. The price of plausibility and coherence is unidimensionality – *i.e.* organizing the structure of the world around one principle. This constructed world in turn provides the ontological stage in which the characters are embedded and the plot unfolds. However, when this same narrative technique is used in NST discourse to extrapolate from current technoscientific abilities to the future, a number of problematic effects are produced.

First, SF literature typically incorporates a historical account, or future history, that explains how the fictional world has come about. It normally contains the period before, during, and after the *novum*. If the narrated world is to be credible, the relationship between the three periods must be one of inevitability. This sense of historical necessity is also reproduced, as I have shown above, when the *novum* structures NST discourse. Thus, as a result of how their respective *nova* have generated the diachronic and synchronic dimensions of their sublime chronotopes, Drexler and Roco and Bainbridge's accounts create a sense of inevitability. However, if the inevitability of these processes are accepted, then there is logically and discursively a rather limited role for ethical reflection or analysis of social implications.

Second, to the extent that the *novum* used to extrapolate a future world is a technoscientific innovation, as is the case in NST discourse, then the extrapolation will take on a technological determinist logic. Technological determinism explains social, cultural, political, and economic change in terms of technoscientific development. However, this logic is a poor operationalization of the dynamics between technoscience and society. The framing of technoscience as the explanatory cause of social phenomena fails to register the complex processes that embed technoscientific practice in specific social, cultural, political, and economic relations. Indeed, as the body of scholarship developing around the social studies of science reveals, technoscience is a social achievement dependent on, for instance, economic rationalities, contests for legitimacy and authority, micro-interactions in the laboratory, social organization, and the development of social networks (Gieryn 1999, Latour & Woolgar 1986). Thus, technoscientific practice relies on the simultaneous production and/or mobilization of social, economic, political, and cultural conditions through which it is invested with legitimacy and effectivity (Latour 1986, Turnbull 2000). The specific ways in which these social processes are negotiated will open up certain developmental pathways while closing off others. It is for this reason that Latour claims that technologies "far from primarily fulfilling a purpose [...] start by exploring heterogeneous universes that nothing, up to that point, could have foreseen and behind which trail new functions" (Latour 2002, p. 250). The extrapolative structure of the *novum* erases the contingencies inherent in technoscientific development by projecting it along a linear developmental path that will most certainly be frustrated. As Brown (2003, p. 4) argues, "In the short term we tend to completely overestimate the practical capabilities of technologies. In the longer-term we tend to get it wrong altogether, with technologies occasionally taking us completely by surprise". This becomes particularly problematic when these developmental paths are invested, as they are within a technological determinist logic made possible by the *novum*, with the ability to resolve all manner of social, cultural, and political problems. Potential non-technological solutions become marginalized and are not pursued.

However, the most fundamental shortcoming of deploying the *novum* as a device for framing discussions on the ethical and social implications of NST is that the *novum* bridges far too many gaps! It not only bridges the technical gap, but also the social and ethical gaps by generating a (fictional) future social world which contains beneficent social implications with only minor ethical complications. As we saw above, the technologies extrapolated from molecular assemblers and NBIC convergence promise a future of prosperity, peace, and physical well-being. Framed in this way, not to promote these technologies and their alleged beneficent social impacts becomes politically negligent if not utterly unethical. However, this momentum towards action obscures the fact that the credibility of the beneficent social implications and the lack of serious ethical conundrums are secured by the narrative structure of the *novum*, not through a critical analysis of social outcomes or serious ethical or normative discussion. Moreover, the *novum* also assigns the social sciences and humanities the function of analyzing and contributing to the management of the social processes necessary to arrive at the proposed future. In this way, they are divested of their potential critical role. For instance, social scientists are asked to analyze public opinion with a view to overcoming public resistance through the effective communication of nano-benefits and promises: *i.e.* by including the public in the political economy of desire and hope generated by the *novum*. They are also asked to aid nano-development by analyzing the mechanisms and procedures which will streamline processes of nano-innovation. In all these contexts, social scientists and humanities scholars are not invited to test the assumptions that underpin the social future generated by the *novum*. Thus, it becomes difficult to envision how a truly critical space is to develop. Moreover, the totalizing utopian vision produced by the *novum* invites similarly generated counter-visions. The latter deploy the structure of the *novum* much as do the former; they differ only in the malevolent logic with which the narrative is invested. Consequently, it is extremely likely that the production of dystopian NST futures may arise not so much from technophobia, fear-mongering, or inadequate knowledge but from the difficulty of criticizing the seemingly impenetrable utopian futures projected through the *novum*. In this context, the most effective critical maneuver is to insert a dystopian virus into a pro-NST program and use its *novum* to assemble a dystopian future.

An understanding of this phenomenon is particularly important because it is the dystopian *novum* that has drawn the attention of popular culture and has, as argued by Marshall (2004), contributed as much to the development of NST as has its utopian counterpart. However, for many of the reasons listed above, in the context of the utopian *novum*, the dystopian register fails as a constructive critique of the social and ethical consequences at stake in NST.

In this study, I have argued that the relation between SF narrative elements and NST is not external but internal. This is due to NST's radical future orientation, which opens up a gap between what is technoscientifically possible today and its inflated promises for the future. I have argued that this gap is bridged by linking the dominant metaphor in NST discourse – the nanotechnoscientist as the *master builder* – to SF narrative techniques used to build future or parallel worlds. I have examined these techniques in detail in two NST texts: Drexler's *Engines of Creation* and Roco and Bainbridge's text on NBIC convergence. I have tried to show how narrative techniques are used in order to extrapolate credible and plausible futures through a synchronic and diachronic re-ontologization of the world.

I have not been concerned with exploring whether this type of narrative process is incompatible with scientific practice. I have, however, identified a number of obstacles that it poses for a critical discussion of ethical and social implications. First, both the sense of inevitability and technological determinism, associated with the *novum*, tend to erode the necessity of analyses of ethical and social implications. Second, I have drawn attention to the fact that the *novum* does more than merely bridge the technical gap; it also bridges the ethical gap by narrating a desirable fictional social world organized around the extrapolated technology. This contains the moral imperative to realize the extrapolated society while simultaneously cloaking the extent to which the plausibility of the extrapolated society is a function of the narrative device of the *novum*. In this context, the role for the humanities and the social sciences is to facilitate the development of the technology rather than to critically engage with it. Finally, I have suggested that the radical immunity to critique that is constructed through the *novum* creates an ideal medium for dystopian counter visions that in turn display many of the same shortcomings in their apocalyptic rendering of NST. I would argue that an understanding of these discursive tendencies must be borne in mind in the attempt to open up a space for a more open and critical analysis of the ethical and social implications of NST.

However, in highlighting how SF narrative elements in NST discourse fail to facilitate effective critical engagement, I am not arguing that SF as a literary genre is not a suitable vehicle for critical reflection on technoscientific developments. It is necessary to be clear about the fact that the existence of SF narrative elements in NST discourse does not make the latter a work of literary SF. What is more, SF as a literary genre is, in fact, better at opening up a space for critical reflection than is the NST discourse described and analyzed in this paper. In other words, ironically literary SF succeeds where NST discourse fails. This is because, as SF writer and critic Samuel R. Delaney argues, "Science fiction is not about the future; it uses the future as a narrative convention to present significant distortions of the present" (1984, p. 47). Similarly, Frederick Jameson (1982) argues that these distortions serve to defamiliarize the present and open up the exploration of alternative social, cultural, and political arrangements. The plausibility of the extrapolated 'future' in SF need only be sufficient to stage the exploration of scientific, political, social, and cultural questions in dramatic form. As a result, SF is less concerned with the 'objective' factors that give rise to a specific future, less concerned with forecasting or describing possible future societies, than [...] with presenting a specific future and discovering what it means to act in specific ways in terms of the belief that those ways of acting are necessary for accepting, rejecting or doubting the principles upon which a particular future social order rests. Thus, the 'future' in SF is only a dramatic device for exploring the present. In contrast, NST discourse confuses the effect of the 'future' produced by the *novum* and its related narrative strategies for the future itself. It confuses the suspended disbelief that is evoked by a world organized around a single principle as a vehicle for a dramatic enactment with foresight. Whereas in SF the extrapolated future is a stepping-stone for critical reflection, in NST discourse the extrapolated future is the endpoint of the reflection. Given the range of techniques, tools, instruments, machines, algorithms, materials, hardware, processes, projects, disciplines, actors, economic interests, and governance agendas that are included under the rubric of NST, it is unlikely that the fields of NST will all develop in unison. More likely than not they will produce varied ethical, legal, and social implications that will have to be monitored and analyzed as they unfold in different social, cultural, political, and economic contexts. An ethical lag is only a problem if we lack the social and political institutions to restrain technoscience when we deem it necessary. A necessary corrective to the unrealistic task of trying to understand the ethical and social implications of NST as if they were one process is to ask ourselves: Does it make sense to group all our macro-technologies in the same way? Moreover, if it is true that the extrapolated future made possible by SF narrative speaks more to the present than the future, we might ask ourselves what are the ethical and social implications of an organization of technoscientific activity that needs to claim such clairvoyance and promise so much to merely function?

Societal Implications of Nanoscience, Nanotechnology and Nanoengineering

Advances in nanoscience and nanotechnology promise to have major implications for health, wealth, and peace in the upcoming decades. Knowledge in this field is growing worldwide, leading to fundamental scientific advances. In turn, this will lead to dramatic changes in the ways that materials, devices, and systems are understood and created. The National Nanotechnology Initiative (NNI) seeks to accelerate that progress and to facilitate its incorporation into beneficial technologies. Among the expected breakthroughs are orders-of-magnitude increases in computer efficiency, human organ restoration using engineered tissue, "designer" materials created from directed assembly of atoms and molecules, and the emergence of entirely new phenomena in chemistry and physics. The study of the societal implications of nanotechnology must be an integral part of the NNI. An interagency effort within the U.S. Government, the NNI supports a broad program of nanoscale research in materials science, physics, chemistry, and biology; it explicitly seeks to create new opportunities for interdisciplinary work. It is balanced across five broad activities: fundamental research; grand challenges; centers and networks of excellence; research infrastructure; and, the ethical, legal, and social implications, including educational and workforce programs. This report outlines some potential areas for research into societal implications of nanotechnology. It has been prepared just as the NNI is commencing, when there is greater opportunity to affect the NNI investment strategy. Research on societal implications will boost the chances for NNI's success and help the nation take advantage of new technology sooner, better, and with greater confidence. Moreover, sober, technically competent research on the interactions between nanotechnology and society will help mute speculative hype and dispel some of the unfounded fears that sometimes accompany dramatic advances in scientific understanding.

Toward this end, the National Science and Technology Council (NSTC), Committee on Technology (CT), Subcommittee on Nanoscale Science, Engineering, and Technology (NSET) the Federal interagency group coordinating the NNI ¾ sponsored a workshop on "Societal Implications of Nanoscience and Nanotechnology." Held September 28-29, 2000 at the National Science Foundation, this workshop brought together nanotechnology researchers, social scientists, and policy makers representing academia, government, and the private sector. Their charge was to: (1) survey current studies on the societal implications of nanotechnology (educational, technological, economic, medical, environmental, ethical, legal, etc.); (2) identify investigative and assessment methods for future studies of societal implications; (3) propose a vision for accomplishing nanotechnology's promise while minimizing undesirable consequences. This report sponsored by NSF incorporates fully the views, opinions and presentations contributed by workshop participants and other leading experts. The NSET report to the NSTC Committee on Technology presents a more concise perspective. The workshop participants offered recommendations to: (a) accelerate the beneficial use of nanotechnology while diminishing the risks, (b) improve research and education, and (c) guide the contributions of key organizations. These recommendations, summarized below, serve as a basis for both the NNI participants and the public to begin addressing societal issues of nanotechnology:

- Make support for social and economic research studies on nanotechnology a high priority.
- Include social science research on the societal implications in the nanotechnology research centers, and consider creation of a distributed research center for social and economic research. Build openness, disclosure, and public participation into the process of developing nanotechnology research and development program direction.
- The National Nanotechnology Coordination Office (NNCO) should establish a mechanism to inform, educate, and involve the public regarding potential impacts of nanotechnology. The NNCO should receive feedback from the nanotechnology community, social scientists, the private sector, and the public with the goals of (a) continuously monitoring the potential societal opportunities and challenges; and (b) providing timely input to responsible organizations.
- Create the knowledge base and institutional infrastructure to evaluate nanotechnology's scientific, technological, and societal impacts and implications from short-term (3 to 5 year), medium-term (5 to 20 year), and long-term (over 20 year) perspectives. This must include interdisciplinary research that incorporates a systems approach (research-technology development-societal impacts), life cycle analysis, and real time monitoring and assessment.
- Educate and train a new generation of scientists and workers skilled in nanoscience and nanotechnology at all levels. Develop specific curricula and programs designed to:
    a. introduce nanoscale concepts into mathematics, science, engineering, and technological education;
    b. include societal implications and ethical sensitivity in the training of nanotechnologists;
    c. produce a sufficient number and variety of well-trained social and economic scientists prepared to work in the nanotechnology area;
    d. develop effective means for giving nanotechnology students an interdisciplinary perspective while strengthening the disciplinary expertise they will need to make maximum professional contributions; and
    e. establish fruitful partnerships between industry and educational institutions to provide nanotechnology students adequate experience with nanoscale fabrication, manipulation, and characterization techniques.

- Encourage professional societies to develop forums and continuing education activities to inform, educate, and involve professionals in nanoscience and nanotechnology.

Over the next 10 to 20 years, nanotechnology will fundamentally transform science, technology, and society. However, to take full advantage of opportunities, the entire scientific and technology community must set broad goals; creatively envision the possibilities for meeting societal needs; and involve all participants, including the general public, in exploiting them.

A revolution is occurring in science and technology, based on the recently developed ability to measure, manipulate and organize matter on the nanoscale — 1 to 100 billionths of a meter. At the nanoscale, physics, chemistry, biology, materials science, and engineering converge toward the same principles and tools. As a result, progress in nanoscience will have very far-reaching impact. The nanoscale is not just another step toward miniaturization, but a qualitatively new scale. The new behavior is dominated by quantum mechanics, material confinement in small structures, large interfacial volume fraction, and other unique properties, phenomena and processes. Many current theories of matter at the microscale have critical lengths of nanometer dimensions. These theories will be inadequate to describe the new phenomena at the nanoscale. As knowledge in nanoscience increases worldwide, there will likely be fundamental scientific advances. In turn, this will lead to dramatic changes in the ways materials, devices, and systems are understood and created. Innovative nanoscale properties and functions will be achieved through the control of matter at its building blocks: atom-byatom, molecule-by-molecule, and nanostructure-by-nanostructure. Nanotechnology will include the integration of these nanoscale structures into larger material components, systems, and architectures. However, within these larger scale systems the control and construction will remain at the nanoscale. Today, nanotechnology is still in its infancy, because only rudimentary nanostructures can be created with some control. However, among the envisioned breakthroughs are orders-of-magnitude increases in computer efficiency, human organ restoration using engineered tissue, "designer" materials created from directed assembly of atoms and molecules, as well as emergence of entirely new phenomena in chemistry and physics. Nanotechnology has captured the imaginations of scientists, engineers and economists not only because of the explosion of discoveries at the nanoscale, but also because of the potential societal implications. A White House letter (from the Office of Science and Technology Policy and Office of Management and Budget) sent in the fall of 2000 to all Federal agencies has placed nanotechnology at the top of the list of emerging fields of research and development in the United States. The National Nanotechnology Initiative was approved by Congress in November 2000, providing a total of $422 million spread over six departments and agencies. Nanotechnology's relevance is underlined by the importance of controlling matter at the nanoscale for healthcare, the environment, sustainability, and almost every industry. There is little doubt that the broader implications of this nanoscience and nanotechnology revolution for society at large will be profound.

The National Nanotechnology Initiative (NNI, http://nano.gov) is a multi-agency effort within the U.S. Government that supports a broad program of Federal nanoscale research in materials, physics, chemistry, and biology. It explicitly seeks to create opportunities for interdisciplinary work integrating these traditional disciplines. The NNI will accelerate the pace of fundamental research in nanoscale science and engineering, creating the knowledge needed to enable technological innovation, training the workforce needed to exploit that knowledge, and providing the manufacturing science base needed for future commercial production. Potential breakthroughs are possible in areas such as materials and manufacturing, medicine and healthcare, environment and energy, biotechnology and agriculture, electronics and information technology, and national security.

The effect of nanotechnology on the health, wealth, and standard of living for people in this century could be at least as significant as the combined influences of microelectronics, medical imaging, computer-aided engineering, and man-made polymers developed in the past century. The NNI is balanced across five broad activities: fundamental research; grand challenges; centers and networks of excellence; research infrastructure; and societal/workforce implications. Under this last activity, nanotechnology's effect on society – legal, ethical, social, economic, and workforce preparation – will be studied to help identify potential concerns and ways to address them. As the NNI is commencing, there is a *rare opportunity to integrate the societal studies and dialogues from the very beginning and to include societal studies as a core part of the NNI investment strategy.*

Research on societal implications will boost the NNI's success and help us to take advantage of the new technology sooner, better, and with greater confidence. Toward this end, the National Science and Technology Council (NSTC), Committee on Technology (CT), Subcommittee on Nanoscale Science, Engineering, and Technology (NSET).... the Federal interagency group that coordinates the NNI, sponsored a workshop on "Societal Implications of Nanoscience and Nanotechnology." Held September 28-29, 2000 at the National Science Foundation, this workshop brought together nanotechnology researchers, social scientists, and policy makers representing academia, government, and the private sector. It had four principal objectives:

- Survey current studies on the societal implications of nanotechnology (educational, technological, economic, medical, environmental, ethical, legal, cultural, etc.).
- Identify investigative and assessment methods for future studies of societal implications.
- Propose a vision and alternative pathways toward that vision integrating short-term (3 to 5 year), medium-term (5 to 20 year), and long-term (more than 20 year) perspectives.

- Recommend areas for research investment and education improvement. This report addresses issues far broader than science and engineering, such as how nanotechnology will change society and the measures to be taken to prepare for these transformations. The conclusions and recommendations in this report will provide a basis for the NNI participants and the public to address future societal implications issues.

Nanoscale science and engineering will lead to better understanding of nature; advances in fundamental research and education; and significant changes in industrial manufacturing, the economy, healthcare, and environmental management and sustainability. Examples of the promise of nanotechnology, with projected total worldwide market size of over $1 trillion annually in 10 to 15 years, include the following:

- *Manufacturing*: The nanometer scale is expected to become a highly efficient length scale for manufacturing once nanoscience provides the understanding and nanoengineering develops the tools. *Materials* with high performance, unique properties and functions will be produced that traditional chemistry could not create. Nanostructured materials and processes are estimated to increase their market impact to about $340 billion per year in the next 10 years (Hitachi Research Institute, personal communication, 2001).
- *Electronics*: Nanotechnology is projected to yield annual production of about $300 billion for the semiconductor industry and about the same amount more for global integrated circuits sales within 10 to 15 years (see R. Doering, page 74-75 of this report).
- *Improved Healthcare*: Nanotechnology will help prolong life, improve its quality, and extend human physical capabilities.
- *Pharmaceuticals*: About half of all production will be dependent on nanotechnology— affecting over $180 billion per year in 10 to 15 years (E. Cooper, Elan/Nanosystems, personal communication, 2000).
- *Chemical Plants*: Nanostructured catalysts have applications in the petroleum and chemical processing industries, with an estimated annual impact of $100 billion in 10 to 15 years (assuming a historical rate of increase of about 10% from $30 billion in 1999; "NNI: The Initiative and Its Implementation Plan," page 84).
- *Transportation*: Nanomaterials and nanoelectronics will yield lighter, faster, and safer vehicles and more durable, reliable, and cost-effective roads, bridges, runways, pipelines, and rail systems. Nanotechnology-enabled aerospace products alone are projected to have an annual market value of about $70 billion in ten years (Hitachi Research Institute, personal communication, 2001).

- *Sustainability*: Nanotechnology will improve agricultural yields for an increased population, provide more economical water filtration and desalination, and enable renewable energy sources such as highly efficient solar energy conversion; it will reduce the need for scarce material resources and diminish pollution for a cleaner environment. For example, in 10 to 15 years, projections indicate that nanotechnology-based lighting advances have the potential to reduce worldwide consumption of energy by more than 10%, reflecting a savings of $100 billion dollars per year and a corresponding reduction of 200 million tons of carbon emissions ("NNI: The Initiative and Its Implementation Plan," page 93).

The study of nanoscale systems promises to lead to fundamentally new advances in science and engineering and in our understanding of biological, environmental, and planetary systems. It also will redirect our scientific approach toward more generic and interdisciplinary research. Nanoscience is at the unexplored frontiers of science and engineering, and it offers one of the most exciting opportunities for innovation in technology. Nanotechnology will provide the capacity to create affordable products with dramatically improved performance. This will come through a basic understanding of ways to control and manipulate matter at the nanometer scale and through the incorporation of nanostructures and nanoprocesses into technological innovations. It will be a center of intense international competition when it lives up to its promise as a generator of technology. Nanotechnology promises to be a dominant force in our society in the coming decades. Commercial inroads in the hard disk, coating, photographic, and pharmaceutical industries have already shown how new scientific breakthroughs at this scale can change production paradigms and revolutionize multibillion-dollar businesses. However, formidable challenges remain in fundamental understanding of systems on this scale before the potential of nanotechnology can be realized. Today, nanotechnology is still in its infancy, and only rudimentary nanostructures can be created with some control. The science of atoms and simple molecules, on one end, and the science of matter from microstructures to larger scales, on the other, are generally established. The remaining size-related challenge is at the nanoscale — roughly between 1 and 100 molecular diameters — where the fundamental properties of materials are determined and can be engineered. A revolution has been occurring in science and technology, based on the recently developed ability to measure, manipulate and organize matter on this scale. Recently discovered organized structures of matter (such as carbo nanotubes, molecular motors, DNA-based assemblies, quantum dots, and molecular switches) and new phenomena (such as giant magnetoresistance, coulomb blockade, and those caused by size confinement) are scientific breakthroughs that merely hint at possible future developments. The nanoscale is not just another step toward miniaturization, but a qualitatively new scale. The new behavior is dominated by quantum mechanics, material confinement in small structures, large interfaces, and other unique properties, phenomena and processes. Many current theories of matter at the microscale have critical lengths of nanometer dimensions; these theories will be inadequate to describe the new phenomena at the nanoscale. Nanoscience will be an essential component in better understanding of nature in the next decades. Important issues include greater interdisciplinary research collaborations, specific education and training, and transition of ideas and people to industry.

The potential benefits of nanotechnology are pervasive, as illustrated in the fields outlined below. Nanotechnology is fundamentally changing the way materials and devices will be produced in the future. The ability to synthesize nanoscale building blocks with precisely controlled size and composition and then to assemble them into larger structures with unique properties and functions will revolutionize materials and manufacturing. Researchers will be able to develop material structures not previously observed in nature, beyond what classical chemistry can offer. Some of the benefits that nanostructuring can bring include lighter, stronger, and programmable materials; reductions in life-cycle costs through lower failure rates; innovative devices based on new principles and architectures; and use of molecular/cluster manufacturing, which takes advantage of assembly at the nanoscale level for a given purpose. The Semiconductor Industry Association (SIA) has developed a roadmap for continued improvements in miniaturization, speed, and power reduction in information processing devices — sensors for signal acquisition, logic devices for processing, storage devices for memory, displays for visualization, and transmission devices for communication. The SIA roadmap projects the future of nanoelectronics and computer technology to approximately 2010 and to 0.1 micron (100 nanometer) structures, just short of fully nanostructured devices. The roadmap ends short of true nanostructured devices because the principles, fabrication methods, and techniques for integrating devices into systems at the nanoscale are generally unknown. New approaches such as chemical and biomolecular computing, and quantum computing making use of nanoscale phenomena and nanostructures, are expected to emerge. The molecular building blocks of life — proteins, nucleic acids, lipids, carbohydrates, and their non-biological mimics — are examples of materials that possess unique properties determined by their size, folding, and patterns at the nanoscale. Biosynthesis and bioprocessing offer fundamentally new ways to manufacture chemicals and pharmaceutical products. Integration of biological building blocks into synthetic materials and devices will allow the combination of biological functions with other desirable materials properties. Imitation of biological systems provides a major area of research in several disciplines. For example, the active area of bio-mimetic chemistry is based on this approach.

Living systems are governed by molecular behavior at the nanometer scale, where chemistry, physics, biology, and computer simulation all now converge. Recent insights into the uses of nanofabricated devices and systems suggest that today's laborious process of genome sequencing and detecting the genes' expression can be made dramatically more efficient through use of nanofabricated surfaces and devices. Expanding our ability to characterize an individual's genetic makeup will revolutionize diagnostics and therapeutics. Beyond facilitating optimal drug usage, nanotechnology can provide new formulations and routes for drug delivery, enormously broadening the drugs' therapeutic potential. Increasing nanotechnological capabilities will also markedly benefit basic studies of cell biology and pathology. As a result of the development of new analytical tools capable of probing the world of the nanometer, it is becoming increasingly possible to characterize the chemical and mechanical properties of cells (including processes such as cell division and locomotion) and to measure properties of single molecules. These capabilities complement (and largely supplant) the ensemble average techniques presently used in the life sciences. Moreover, biocompatible, high-performance materials will result from the ability to control their nanostructure. Artificial inorganic and organic nanoscale materials can be introduced into cells to play roles in diagnostics (e.g., quantum dots in visualization), but also potentially as active components. Finally, nanotechnologyenabled increases in computational power will permit the characterization of macromolecular networks in realistic environments. Such simulations will be essential for developing biocompatible implants and for studying the drug discovery process. An open issue is how the healthcare system would change with such large changes in medical technology.

Nanotechnology will lead to dramatic changes in the use of natural resources, energy, and water, as outlined in the following paragraphs. Waste and pollution will be minimized. Moreover, new technologies will allow recovery and reuse of materials, energy, and water.

Environment: Nanoscience and engineering could significantly affect molecular understanding of nanoscale processes that take place in the environment; the generation and remediation of environmental problems through control of emissions; the development of new "green" technologies that minimize the production of undesirable by-products; and the remediation of existing waste sites and streams. Nanotechnology also will afford the removal of the smallest contaminants from water supplies (less than 200 nanometers) and air (under 20 nanometers) and the continuous measurement and mitigation of pollution in large areas. In order to hasten the integrated understanding of the environmental role of nanoscale phenomena, scientists and engineers studying the fundamental properties of nanostructures will need to work together with those attempting to understand complex processes in the environment. Model nanostructures can be studied, but in all cases the research must be justified by its connection to naturally occurring systems or to environmentally beneficial uses. Environments for investigations are not limited and might include terrestrial locations such as acid mines, subsurface aquifers, or polar environments. *Energy:* Nanotechnology has the potential to significantly impact energy efficiency, storage, and production. Several new technologies that utilize the power of nanostructuring, but developed without benefit of the new nanoscale analytical capabilities, illustrate this potential:

- Increasing the efficiency of converting solar energy into useful forms.

- High efficiency fuel cells, including hydrogen storage in nanotubes.
- A long-term research program in the chemical industry on the use of crystalline materials as catalyst supports has yielded catalysts with well-defined pore sizes in the range of 1 nanometer to reduce energy consumption and waste; their use is now the basis of an industry that exceeds $30 billion a year ("NNI: The Initiative and Its Implementation Plan,").
- Developed by the oil industry, the ordered mesoporous material MCM-41 (known also as "self-assembled monolayers on mesoporous supports," SAMMS), with pore sizes in the range of 10-100 nanometers, is now widely used for the removal of ultrafine contaminants (see work performed at Pacific Northwest National Laboratory in *Nanotechnology Research Directions,* Kluwer Academic Publishers, 2000, pp. 216-218).
- Several chemical manufacturing companies are developing a nanoparticle-reinforced polymeric material that can replace structural metallic components in automobiles; widespread use of those nanocomposites could lead to a reduction of 1.5 billion liters of gasoline consumption over the life of one year's production of vehicles, thereby reducing carbon dioxide emissions annually by more than 5 billion kilograms ("NNI: The Initiative and Its Implementation Plan,").
- Significant changes in lighting technologies are expected in the next ten years. Semiconductors used in the preparation of light emitting diodes (LEDs) for lighting can increasingly be sculpted on nanoscale dimensions. In the United States, roughly 20% of all electricity is consumed for lighting, including both incandescent and fluorescent lights. In 10 to 15 years, projections indicate that such nanotechnologybased lighting advances have the potential to reduce worldwide consumption of energy by more than 10%, reflecting a savings of $100 billion dollars per year and a corresponding reduction of 200 million tons of carbon emissions ("NNI: The Initiative and Its Implementation Plan," pages 92-93).
- The replacement of carbon black in tires by nanometer-scale particles of inorganic clays and polymers is a new technology that is leading to the production of environmentally friendly, wear-resistant tires.

Agriculture: Nanotechnology will contribute directly to advancements in agriculture in a number of ways: (1) molecularly engineered biodegradable chemicals for nourishing plants and protecting against insects; (2) genetic improvement for animals and plants; (3) delivery of genes and drugs to animals; and (4) nano-array-based technologies for DNA testing, which, for example, will allow a scientist to know which genes are expressed in a plant when it is exposed to salt or drought stress. The application of nanotechnology in agriculture has only begun to be appreciated.

Water: Global population is increasing while fresh water supplies are decreasing. The United Nations predicts that by the year 2025 that 48 countries will be short of fresh water accounting for 32% of the world's population ("NNI: The Initiative and Its Implementation Plan", page 95). Water purification and desalinization are some of the focus areas of preventative defense and environmental security since they can meet future water demands globally. Consumptive water use has been increasing twice as fast as the population and the resulting shortages have been worsened by contamination. Nanotechnology-based devices for water desalinization have been designed to desalt sea water using at least 10 times less energy than state-of–the art reverse osmosis and at least 100 times less energy than distillation. The critical experiments underpinning these estimations are underway now. This energy-efficient process is possible by fabricating of very high surface area electrodes that are electrically conductive using aligned carbon nanotubes, and by other innovations in the system design.

The stringent fuel constraints for lifting payloads into earth orbit and beyond, and the desire to send spacecraft away from the sun for extended missions (where solar power would be greatly diminished) compel continued reduction in size, weight, and power consumption of payloads. Nanostructured materials and devices promise solutions to these challenges. Nanostructuring is also critical to the design and manufacture of lightweight, high-strength, thermally stable materials for aircraft, rockets, space stations, and planetary/solar exploratory platforms. The augmented utilization of miniaturized, highly automated systems will also lead to dramatic improvements in manufacturing technology. Moreover, the low-gravity, high-vacuum space environment may aid the development of nanostructures and nanoscale systems that cannot be created on Earth.

Defense applications include (1) continued information dominance through advanced nanoelectronics, identified as an important capability for the military; (2) more sophisticated virtual reality systems based on nanostructured electronics that enable more affordable, effective training; (3) increased use of enhanced automation and robotics to offset reductions in military manpower, reduce risks to troops, and improve vehicle performance; (4) achievement of the higher performance (lighter weight, higher strength) needed in military platforms while providing diminished failure rates and lower life-cycle costs; (5) needed improvements in chemical/biological/nuclear sensing and in casualty care; (6) design improvements in systems used for nuclear non-proliferation monitoring and management; and (7) combined nanomechanical and micromechanical devices for control of nuclear defense systems. In many cases economic and military opportunities are considered to be complementary. Strong applications of nanotechnology in other areas would provide support for national security in the long term, and vice versa.

Since economists have not yet really begun research on nanotechnology, their insights are somewhat tentative and based on experience with earlier technologies. A common paradigm is that new applications will be initially more costly than existing technologies, but will achieve better performance. However, completely new technologies may be cheaper, such as chemical manufacturing to mass produce nanoelectronic circuits as opposed to current methods using lithography in microelectronics. Overall, nanotechnology will offer substantial advantages, being smaller, faster, stronger, safer, and more reliable. At the same time, it will require investments in new production facilities and in a host of ancillary industries supplying the raw materials, components, and manufacturing machines. Because it will take time to achieve economies of scale and to develop the most efficient fabrication methods, costs are likely to be relatively high in the beginning. For this reason, nanotechnology-based goods and services will probably be introduced earlier in those markets where performance characteristics are especially important and price is a secondary consideration. Examples are medical applications and space exploration. The experience gained will reduce technical and production uncertainties and prepare these technologies for deployment into the market place. Similarly, in the private sector, technology transfer is likely to occur from performance-oriented areas (such as medicine) to price-oriented ones (such as agriculture). As a given technology matures, its cost may decline, leading to greater penetration of the market even where performance is not decisive. The displacement of an old technology by a new one tends to be both slow and incomplete. Displacement of older methods will accelerate to the extent that nanotechnology extends its technical range and perhaps lowers its relative price. However, nanotechnology also is likely to stimulate innovations in older technologies that make them better able to compete — an ironic but potentially beneficial secondorder effect. The diffusion and impact of nanotechnology will be partly a function of the development of complementary technologies and of a network of users. Whole new industries may have to be developed — along with the trained scientists and technicians to staff them. There may be many obstacles along this road that ordinary market processes cannot easily overcome. An important role for the government will be to invest in the long-term, high-risk, high-gain research needed to create these new industries and to ensure that they are consistent with broader societal objectives. Federal support of the nanotechnology initiative is necessary to enable the United States to take advantage of this strategic technology and remain competitive in the global marketplace well into the future. Focused research programs on nanotechnology have been initiated in other industrialized countries. Currently, the United States has a lead on synthesis, chemicals, and biological aspects; it lags in research on nanodevices, production of nano-instruments, ultra-precision engineering, ceramics, and other structural materials. Japan has an advantage in nanodevices and consolidated nanostructures; Europe is strong in dispersions, coatings, and new instrumentation.

Technological Determinism in Nanoscience

Is (nano)technology a product of society, or is society a product of technology? Do social groups construct what counts as 'progress' in the development of a technology, or do artifacts and systems evolve according to their own, internal rules? These are the questions that once sparked vigorous debate over 'technological determinism'. Yet in the past few years philosophers, historians, and sociologists of technology have largely steered away from these thorny issues. Stark versions of determinist thinking, such as Lynn White's (1962) claim that feudalism was a product of the stirrup and the heavy plow, or, for that matter, Marx's (1963/1847) remark that "the hand mill gives you society with the feudal lord; the steam mill society with the industrial capitalist" today seem too oversimplified even to provoke scholarly discussion. As one of the last important contributions to this debate, the edited volume *Does Technology Drive History?* (Smith & Marx 1994), answered its eponymous question – 'not really'. One of the problems with sustaining analysis of technological determinism is that there is little agreement about what it is. Indeed, in the decade between 1985 and 1995, there briefly flourished a cottage industry devoted to splicing apart the various threads of determinist thought, giving them names, and associating them with different schools of philosophy and history. As Bruce Bimber (1994) pointed out in a landmark article, technological determinism "exists in enough different incarnations that the label can easily be attached to a range of views". Within this range, one can find a spectrum from 'strong' to 'weak' determinism – for some, technology may be *the* driving force of social change, while for others (most notably Thomas Hughes) a technological system may seem to have an autonomous, extra-social 'momentum' (Hughes 1983, 1994) that drives social change only because society itself provides the soil to grow networks of power, standards, institutions, and artifacts that entrench the system by enrolling vast numbers of stakeholders.

Thus, much of the attraction of determinist representations of technology's development and effect on society may lie in the interpretive flexibility of applying both 'technology' and 'determinism' to any particular case. Yet, though marking out the different senses latent within technological determinism was an important project, it has tended to end rather than provoke debate. For the purposes of this article, therefore, I wish to point to conceptual territory that may lie beyond the parsing of definitions. To do so, I will rely on a two-handed definition of technological determinism borrowed from Bijker (1995). In Bijker's summary, technological determinism encompasses *both* the idea that technological development proceeds via an autonomous, internal logic (a logic determined only by a unidirectional calculus of engineering considerations, rather than a dense weave of contradictory aims that are both 'social' and 'technical') *and* the idea that technology determines the social organization of a society (and therefore pushes rather than pulls societal change). As Bijker points out, though, the two notions are intertwined. Because technology is seen as prior to, rather than an upshot of, society, it is easy to think of technological choices as having their own, pure logic; and because technological changes are thought to accumulate under their own power (and simultaneously provide the motive force for societal change), it is almost axiomatic (at least in North America and many other Western societies) that technical development can be used as a (or usually *the*) yardstick in measuring how 'advanced' a culture is. The advantage of this particular definition is that it highlights elements of technological determinism within both mainstream, popular ideology, and academic philosophy and history. To be sure, outside of technology studies circles determinist talk is still alive and well. Popular representations of technology, as well as policy statements by proponents and opponents of particular artifacts and systems, paint technologies as possessing autonomy, as developing along ineluctable pathways, and as being the core around which society is structured and measured. Indeed, this provides its own fodder for analytical debate as historians and sociologists examine how advocates' and opponents' *representations* of technology as autonomous shape both the design of artifacts and the social order surrounding them in ways that recursively give the technology a deterministic social reality. As historians Gabrielle Hecht and Michael Thad Allen argue,

> [I]nstead of continuing to ask 'Does technology drive history?' we should ask questions such as 'When or why do historical actors believe or argue that technology drives history?' Addressing such questions leads us to view technological determinism – and other beliefs about the relationships between technology and social change – as political practices. [Hecht & Allen 2001, p. 14-15]

Though determinist talk of all stripes – strong and weak, nuanced and simple – is ubiquitous, it is often easiest to capture and analyze pronouncements made about *emerging* technologies. This may seem counterintuitive; after all, emerging technologies are thinly connected to networks of people and institutions, are easily reconstrued as new participants have their say, and continually face the specter of failure and disappearance. Unlike many entrenched technologies, emergent systems usually spawn a variety of contradictory voices. Yet, though these voices differ, they still often reinforce a technologically determinist worldview by laying out a determined path for the technology's development and a means by which the technology will ineluctably reshape society. The strong association between emergent technologies and determinist talk seems less paradoxical, though, if we see such statements as *performative*, rather than reflective, of a determinist viewpoint. Technology's advocates build networks of people and institutions *through* determinist talk and action, and in doing so they conjure up the thick social ties that make such determinism plausible. Few of today's emerging technologies fit this model better than nanotechnology. Nano's proponents, in particular, are not shy about saying that current research will inevitably generate a brave new world that will look completely different from pre-nano society. In engaging analytically with such promises, scholars of science and technology have a tremendous opportunity. Nano represents a scientific and technological movement in the making (or, perhaps, unmaking). Nano should be viewed as an exquisite field site for testing our ideas about how people generate knowledge and artifacts; how they integrate new technologies into their practices and organize themselves around new kinds of artifacts; and, indeed, how they use emerging technologies to push the limits of human instrumentality. For these reasons, nano is fertile ground for sharpening historical, philosophical, and sociological analysis of technological determinism. Yet, nano, as currently constituted, also displays a number of wrinkles on classical determinism that make it interesting as more than a mere test case. Most fascinating and analytically useful is its proponents' cultivation of it as a simultaneously scientific and technological endeavor. Nanoists routinely mix scientific and technological registers in their talk; and in their practice, they devise experiments that can easily be construed both/either as generating interesting scientific knowledge and/or useful technological artifacts. Interestingly, nanoists often project this synthesis far back into the past and forward into the future by, for instance, saying that nanoscience has been gathering steam (perhaps unnoticed) for a very long time in the guise of research in fields such as chemistry and materials science, or that nanotechnology has long been present in practices such as glass-making and blacksmithing where craft knowledge can produce striking nanoscale effects. Moreover, they say, nature (or 'biology') has been doing nanotechnology for billions of years; every virus, bacterium, and cell is a nanomachine of enormous complexity. Indeed, it is around this point that some nanoists invoke a complex but strong form of determinism. After all, nature's nano-achievements show us that nanomachines are possible, and nature's version of nano has completely restructured the earth and produced human life, culture, and consciousness. The progress of science, they say, means that it will inevitably be possible for us to understand and mimic nature's nanomachines; once we have done so, our own nanomachines will develop in a way determined by biology, chemistry, and engineering design; and as they do develop, our inventions cannot help but revolutionize our world just as much as nature's nanobots did.

Thus, nano – and the determinist rhetoric that surrounds it – plays with and synthesizes distinctions between science and technology in interesting ways. This makes nano ripe for the kind of analysis that extends almost a century-long tradition of using the philosophy, history, and sociology of science and technology to cast light on each other. The strands of this tradition that I will draw on here begin with Dewey and Heidegger, and pass through Bachelard and Wittgenstein and Kuhn, but have taken on many colors with the advent of the science and technology studies literature in the late 1970s. Indeed, today scholars as diverse as Don Ihde, Trevor Pinch, Gabrielle Hecht, Ian Hacking, Peter Galison, and Bruno Latour have used our understanding of science to sharpen analysis of technology *and* vice versa. Of these post-Kuhnian literatures, this article draws most heavily on the social construction of technology (or SCOT) model associated with Bijker and Pinch (1987). SCOT is particularly appropriate here since the model cut its teeth in the 1980s on the debates over technological determinism. In particular, by showing that there is 'interpretive flexibility' in the way engineering choices are made (and therefore no wholly autonomous logic of design is possible) and that technologies are continually reshaped and reinterpreted as new social groups become relevant to them (and therefore technology cannot straightforwardly 'impact' social organization), SCOT countered most (strong) determinist arguments and contributed to the shift away from the debate on technological determinism.

The lessons of SCOT and other post-Kuhnian literatures are many, but a few are key in examining the role of determinism in the relationship between nanotechnology and its constituent communities of practice. First, whatever the metaphysical nature of reality, the sciences as they are actually constituted deal almost exclusively not with the 'real world' but with a world that has been appropriated for human action. That is, scientists engage with a world that they manufacture to be more amenable to the generation of knowledge, and then they learn what they can about *that* world. They clean this reconstituted world, they filter it, they abstract it, they mold it into model systems, and they stimulate it to produce and be populated by some entities rather than others. Thus, the scientific world is inherently technological, and scientists create knowledge by piecing together generative relationships between different *made* objects – microscopes, accelerators, electrons, lab rats, *etc.* Hence, different regions of science and engineering have quite different epistemic materials and therefore quite different practices and bodies of knowledge. Different disciplines and subdisciplines have a certain autonomy because of their arcane knowledge of how to tame the world in their peculiar way and learn something about it. Thus, the knowledge of one science should not be seen as reducible to the knowledge of another, nor should the work of engineers in creating a world that is amenable to their technological expertise be seen as a mere 'application' of any scientific discipline's body of knowledge. From this also follows the Kuhnian point that these crafted worlds make scientific progress difficult to measure. Disciplines change their world-creating practices over time, and hence the knowledge of one era relates to a set of entities that is, in some sense, incommensurable to the knowledge of another era. By the same token, this line of reasoning problematizes notions of technological determinism. Fine-grained studies of scientific practice show that new laboratory technologies do not fit unproblematically into ongoing research communities; rather, the technologies have to be reworked and made compatible with the community's practices. Thus, the design of a technology does not determine its use, and there is no determined relation between a research community's organization and the technologies it employs.

Yet, technologies *can* travel between communities, different disciplines clearly *can* communicate with each other, and different kinds of practitioners *can* harmonize their practice. What is required for this are bits of crafted world – 'boundary objects' (Star & Griesemer 1989) – that can be passed as tokens and made the focus of work that is sufficiently, but not completely, harmonized between different kinds of practitioners. Again, this way of looking at things brings out many of the most conspicuous characteristics of nanotechnology. Like any of the traditional big scientific disciplines, nanotechnology is a community of communities – it contains an overlapping yet mixed bag of surface scientists, probe microscopists, semiconductor physicists, supramolecular chemists, molecular biologists, computer scientists, electrical engineers, materials scientists, UV and electron lithographers, micro-electromechanical systems experts, and so on. Unlike the traditional disciplines, though, there has been little attempt to claim, so far, that the expertise of the constituent parts of nanotechnology is fully commensurable. Policy specialists, practicing scientists and engineers, and sociologists and philosophers of science and technology have all had tremendous difficulty even arriving at a coherent *definition* of nanotechnology, much less a common jargon for all of the knowledge created by self-described nanotechnologists.

Several of the constituent communities of nanotechnology are drawn from the engineering sciences – materials science, electrical engineering, mechanical engineering, fluid dynamics, computer science, MEMS, *etc.* Since the 1970s, these subdisciplines have spawned their own literature in the science and technology studies tradition, a literature that has consistently engaged and critiqued technological determinism in ways that will be helpful in understanding nanotechnology. Scholars such as Ed Layton, Ed Constant, Walter Vincenti, Ron Kline, Eda Kranakis, and Thomas Hughes have shown that engineering has its own practices, its own kinds of instrumentation, theories, and heuristics, and a body of knowledge that cannot simply be reduced to physics. Moreover, these scholars have demonstrated that rhetorical repertoires of 'science' and 'technology' or of 'pure' and 'applied' science are historically situated and closely connected to struggles over the disciplinary identity and autonomy of the engineering sciences (Kline 1995, 2000). The historical sensibility these authors provide is useful in considering nanotechnology as merely the latest in a long line of attempts to provide a heuristic and organizational umbrella over different patches of the engineering disciplines, and the rhetoric of nanoists as performative in the construction of their umbrella. Nano also has a strong constituency from scientific subdisciplines, especially those currently housed in traditional chemistry departments. Even before Dalton and atomism, chemists knew their discipline dealt with very small objects, and modern chemistry is the birthplace of canonically nanotechnological 'artifacts' such as the nanotube, the buckyball, and the DNA computer. In the past, because of the reductionist bent of certain kinds of logical empiricism, and because of the social prestige of physics, chemistry was often overlooked by sociologists and philosophers; there were very good histories of chemistry, such as the classic Guerlac (1961), but little exploration of how the epistemics and social practice of chemistry differed from physics. As with the engineering sciences, though, there is now a burgeoning literature showing that chemists have their own kind of relationship to instrumentation, that they treat issues of purity and contamination in their own (epistemically significant) way, and that they have a different kind of bodily engagement with their experiments and representations than other scientists. Most importantly, this literature draws out the sense in which chemistry is the consummate science of *making* 'epistemic things' – materials that provide a stage for ongoing experimental work and that yield up some small part of the world for scrutiny. The purview of chemistry *is* the making of molecules, integrated with the equipment, concepts, and processes that allow chemists to simultaneously generate knowledge and nanoscale objects.

Engineers and chemists both bring a thing-making orientation to nanotechnology. What is perhaps new for chemists, though, is the idea that the *epistemic* materials they are making should be construed primarily as *technological* artifacts (or parts thereof). It is this process of recasting that has provided much of the hype of nanotechnology, as well as some of the internal frictions of the nano community. It is not immediately obvious in what sense molecules or supramolecular assemblies should be viewed as technological artifacts; and those who have made that leap have sometimes attracted criticism for doing so. This is true of no one more than Eric Drexler, the popularizer of the term 'nanotechnology' and one of the most influential visionaries of the field. It is worthwhile examining Drexler's rhetoric, and his evolving place in the nano community, to understand how this synthesis of chemistry and engineering can yield new forms of technological determinism. Interestingly, Drexler's background is as a futurist, rather than as a practitioner of any of nanotechnology's constituent communities. During his undergraduate education at MIT in the late 1970s, he became a protégé of space travel visionary Gerard K. O'Neill and artificial intelligence futurist Marvin Minsky. At the same time, he kept close track of the dramatic changes in molecular biology and genetic engineering of the day and began developing his own ideas about how artificially engineered biomolecules could be used to further his mentors' dreams of space exploration and artificial intelligence. By 1981 he had begun publishing his vision under the label of 'nanotechnology' – a vision in which very small 'assemblers', modeled on biological machines (cells, ribosomes, viruses, *etc.*), could reconstitute raw materials into almost any physically possible artifact (Drexler 1981).

In 1986, Drexler and his wife, Christine Peterson, along with a group of like-minded friends, moved to Palo Alto to found the Foresight Institute, an organization dedicated to predicting and planning for the dramatic changes caused by nanotechnology. At this time, Drexler formed personal and intellectual links with other futurists in the Bay Area, particularly Stewart Brand, founder of the *Whole Earth Catalog*, that helped legitimate Drexler's project and provided a model for the niche he began to fill. This tradition of futurism, with roots going back through Werner von Braun and Arthur C. Clarke to at least as far back as H.G. Wells and Jules Verne, has left a profound imprint on nanotechnology. All nanotechnologists – whether supporters or critics of Drexler – must deal with his legacy, even if he can no longer fully control his bequest; and that legacy bears the mark of the futurist community.

This futurist inheritance ought to spur particular kinds of analytical discussions of nanotechnology. Historians and sociologists, for instance, will have to place Drexler and nanotechnology in this visionary tradition and delineate the linkages between different kinds of futurism latent in his work. Philosophers, meanwhile, should investigate the unusual time horizons that govern nanotechnological work. It may be useful, for example, to develop a concept of 'presentist' and 'non-presentist' disciplines. Physics and chemistry, for instance, have a more or less presentist orientation. Results generated in the now are drafted into a body of knowledge that is conceived as applying regardless of date. Except for sub-fields like cosmology and geochemistry, the past and future are conceived as being essentially like the present, so that the present is the only arena of experimentation that matters. Nanotechnology, on the other hand, seems decidedly non-presentist. Most traditional disciplines restrict their focus to the materials and instruments (the 'made world') presently available to them. As Drexler and other nano elites often point out, though, nanotechnology came of age at the same time as widespread, powerful computing. Thus, nanotechnology is intensely grounded in computer simulations, and much of the 'made world' of nano has a virtual, yet-to-be-realized quality (Lenhard 2004). Nanotechnologists work as much in this future world as in the present. Drexler himself nicely sums up this orientation and its debt to the futurist tradition:

> Scientists are encouraged by their colleagues and their training to focus on ideas that can be tested with available apparatus. The resulting short-term focus often serves science well: it keeps scientists from wandering off into foggy worlds of untested fantasy [...] [E]ngineers share similar leanings toward the short term [...] [S]cientists refuse to predict future scientific knowledge, and seldom discuss future engineering developments. Engineers do project future developments, but seldom discuss any not based on present abilities. Yet this leaves a crucial gap: what of engineering developments firmly based on *present science* but awaiting *future abilities*? [...] Imagine a line of development which involves using existing tools to build new tools, then using *those* tools to build novel hardware (perhaps including yet another generation of tools) [...] Recent history illustrates this pattern. Few engineers considered building space stations before rockets reached orbit [...] Similarly, few mathematicians and engineers studied the possibilities of computation until computers were built. [Drexler 1990, pp. 46-7, italics in original]

Currently, nano experiments often yield *knowledge* that is siphoned into the experimenter's home discipline (physics, chemistry, *etc.*); but the epistemic *value* of the experiment for nano itself is that it provides a 'proof of concept' for some process or mechanism that – in the future – can be integrated into a more complex nanomachine. That is, nano results are framed in terms of how they contribute to an envisioned path of engineering evolution that necessitates small, cumulative design advances along the way. To flesh out the roots of nanotechnology's non-presentist orientation, it is worth doing a close reading of Drexler's first popular book, *Engines of Creation: The Coming Era of Nanotechnology*. This is the book that first pushed nanotechnology into the public consciousness, and, through its influence on policy makers, science fiction writers, journalists, and practicing scientists, continues to shape the practice of the field. It lays out Drexler's vision of atomically-precise technology, then jumps from one staid futurist topic to another (space travel, artificial intelligence, immortality, new media) demonstrating that nanotechnology will revolutionize each of them. The basic points on which the book's argument hinges are unabashedly determinist and non-presentist: nanotechnology is inevitable, and when it comes it will change everything.

> Assemblers will take years to emerge, but their emergence seems almost inevitable: Though the path to assemblers has many steps, each step will bring the next in reach, and each will bring immediate rewards. The first steps have already been taken, under the names of 'genetic engineering' and 'biotechnology' [...] Barring worldwide destruction or worldwide controls, the technology race will continue whether we wish it or not [...] To have any hope of understanding our future, we must understand the consequences of assemblers, disassemblers, and nanocomputers. They promise to bring changes as profound as the industrial revolution, antibiotics, and nuclear weapons all rolled up in one massive breakthrough. To understand a future of such profound change, it makes sense to seek principles of change that have survived the greatest upheavals of the past. [Drexler 1990, p. 20]

The reason nanotechnology is inevitable is that we have a model for how to proceed: natural, biological nanoscale 'machines'. According to Drexler, we are on the verge not only of understanding these biomachines, but of mimicking them:

> [S]imple molecules make up passive substances. More complex patterns make up the active nanomachines of living cells. Biochemists already work with these machines, which are chiefly made of protein, the main engineering material of living cells [...] [P]rotein machines are unusually flexible. But like all machines, they have parts of different shapes and sizes that do useful work. All machines use clumps of atoms as parts. Protein machines use very small clumps. Biochemists dream of designing and building such devices, but there are difficulties to be overcome [...] When they combine molecules in various sequences, they have only limited control over how the molecules join. When biochemists need complex molecular machines, they still have to borrow them from cells. Nevertheless, advanced molecular machines will eventually let them build nanocircuits and nanomachines as easily and directly as engineers now build microcircuits or washing machines. Then progress will become swift and dramatic. [Drexler 1990, p. 6]

Why will progress be swift and dramatic? In *Engines of Creation* and his more technical sequel, *Nanosystems: Molecular Machinery, Manufacturing, and Computation*, Drexler makes an exact, systematic analogy between biological 'nanomachines' (and their parts) and macroscale engineering artifacts (and their parts). In Drexler's view, nanotechnology will inevitably progress by translating the principles of macroscale engineering into their nanoscale equivalents:

> The similarities between nanomachines and macromachines are pervasive and fundamental. At the analytical level, systems of both kinds can be described by applying classical mechanics to objects that occupy space, exclude other objects from that space, and resist deformation. At the design level, systems of both

> kinds must apply forces, guide motions, limit friction, and so forth [...] Because functions at the system level can usually be implemented in many different ways at the component level, the parallels between macro and nanoscale systems can be even stronger than those between their components. Accordingly, many of the lessons of macroscale mechanical engineering can be applied directly. When nanomechanical designs are drawn at a scale and resolution that omits atomic detail, they can be almost indistinguishable (save for dimensioning labels) from designs for macromachines. [Drexler 1992, pp. 315-6]

Reading Drexler's technical work can be a bit like flipping through Diderot and d'Alembert's *Encyclopedie* – he introduces all the classical machines and their parts, and then offers simulations of their nano-equivalents. Note, for instance, the sub-headings of sections 10.5 through 10.7 in *Nanosystems*, in which he describes a series of simple machines made from small numbers of atoms: 'Nuts and Screws', 'Rods', 'Springs', 'Bearings', 'Spur Gears', 'Helical Gears', 'Rack-and-Pinion Gears and Roller Bearings', 'Bevel Gears', 'Worm Gears', 'Belt-and-Roller Systems', 'Cams', and 'Planetary Gear Systems'. In articulating his argument, Drexler relies on a form of technological determinism that Wiebe Bijker (1995b) calls the 'autonomous logic of technological development' variant. That is, Drexler sees nanotechnology unfolding in a stepwise, progressive fashion, where each step is related to the next by an inherent design rationale – a rationale that can be made visible through the analogy to macroscale technological systems built up from individual machines that are themselves composed of simpler components. Note, though, how Drexler's vision for the evolution of nano-design relies on an *historical* analogy to the evolution of macro-design. The quaintly Enlightenment character of Drexler's nanomachines is symptomatic of a pervasive, forward- *and* backward-looking non-presentism in his writing. Hardly a page goes by in *Engines of Creation* without a pronouncement about a myriad of pasts. Sometimes, Drexler presents nanotechnology as a radical break with these pasts:

> [M]odern technology builds on an ancient tradition. Thirty thousand years ago, chipping flint was the high technology of the day. Our ancestors grasped stones containing trillions of trillions of atoms and removed chips containing billions of trillions of atoms to make their axheads [...] The ancient style of technology that led from flint chips to silicon chips handles atoms and molecules in bulk; call it *bulk technology*. The new technology will handle individual atoms and molecules with control and precision; call it *molecular technology*. It will change our world in more ways than we can imagine. [Drexler 1990, p. 4, italics in original]

At other times, Drexler offers views on a past that can be mined for lessons in organizing this new molecular technology. Indeed, a central – and often overlooked – part of Drexler's argument is that nanotechnology has a long, long past that demonstrates the inevitable success of efforts in the present:

> Simple molecular devices combine to form systems resembling industrial machines. In the 1950s engineers developed machine tools that cut metal under the control of a punched paper tape. A century and a half earlier, Joseph-Marie Jacquard had built a loom that wove complex patterns under the control of a chain of punched cards. Yet over three billion years before Jacquard, cells had developed the machinery of the ribosome. Ribosomes are proof that nanomachines built of protein and RNA can be programmed to build complex molecules. [Drexler 1990, p. 8]

Ribosomes are 'proof', and a three billion year old proof at that; here and elsewhere, we see that Drexler's nanotechnology possesses an epistemic frame in which 'proof' is not a demonstration of certain knowledge about the present state of nature, but rather a performance of a new kind of relationship between how things once were and how they will, inevitably, come to be.

Though he made the term 'nanotechnology' current, and continues to profoundly influence the debates surrounding it, Drexler is by no means the only voice for the field. Indeed, at least since the founding of the US National Nanotechnology Initiative (NNI) in 2000, Drexler's perspective has continually faced challenges from all of the other stakeholders in the enterprise. Those who seek to make nanotechnology a coherent, well-funded, publicly-supported discipline in the present have tried hard in the past few years to separate the field from its futurist past. Above all, this means separating it from Drexler, and both prominent and ordinary nanotechnologists have participated in his ritual expulsion in an attempt to mainstream their discipline. Debates between Drexler and his critics often center on his non-presentist, determinist reasoning. Some of his critics find his analogy between humanly engineered nanomachines and biological 'machines' unconvincing; therefore, they do not accept the three billion year old proof that molecular assemblers can work; hence, they do not see nanotechnology traveling down the path of progressively more complex nanomachines that Drexler lays out; and, therefore, they find Drexler's vision of how the world will be transformed by nano unbelievable. These objections to Drexler's framing of a non-presentist, determinist nanotechnology can be seen in his well-known debate with Nobel Prize-winning chemist Richard Smalley. The crux of the debate is the so-called 'fat fingers, sticky fingers' issue – the idea that molecular assemblers will be unable to pick up and precisely release atoms (as Drexler envisions) because chemical bonds are too 'sticky' and because any assembler will be unable to choose exactly which of many atoms it will interact with (its fingers are too 'fat'). We will return to the image of nano-fingers and nano-limbs later in this paper, but for now it is important to note that Smalley's critique centers on the conspicuous features of Drexler's reasoning that I have outlined above:

> You [*i.e.* Drexler] write that the assembler will use something 'like enzymes and ribosomes' [...] But where does the enzyme or ribosome entity come from in your vision of a self-replicating nanobot? Is there a living cell somewhere

> inside the nanobot that churns these out? There must be liquid water present somewhere inside, and all the nutrients necessary for life [...] Biology is wondrous in the vast diversity of what it can build, but it can't make a crystal of silicon, or steel, or copper, or aluminum, or titanium, or virtually any of the key materials on which modern technology is built [...] If the nanobot is restricted to be a water-based life form, since this is the only way its molecular assembly tools will work, then there is a long list of vulnerabilities and limitations to what it can do. If it is a non-water-based life-form, then there is a vast area of chemistry that has eluded us for centuries [...] You cannot make precise chemistry occur as desired between two molecular objects with simple mechanical motion along a few degrees of freedom in the assembler-fixed frame of reference. [Baum *et al.* 2003, pp. 39-40]

Yet these key modules of Drexler's argument appear again and again in nano discussions, from supporters and critics alike. For example, his likening of genetic material to a computer punch tape that 'instructs' organelles (like some miniscule Turing machine) taps into a broad usage that has old roots in fields such as postwar genetics, information theory, and cybernetics that have branched into nanotechnology. Drexler's more general, and exact, analogy between those nanomachines that are old and biological and those that are new and artificial is also ubiquitous in nano circles.

> Imagine a motor measuring a few hundredths of a thousandth of a millimeter, running on and on. Or a data storage device squeezing the equivalent of five 'high-density' floppy disks into a thousandth of a millimeter [...] We are talking about complicated and highly efficient machines having a size of only a few millionths of a millimeter. Unbelievable? Not at all, for evolution solved these problems more than a billion years ago. The motor mentioned above is already in existence – it is a system mainly consisting of the proteins actin and myosin, and serves to power our muscles. The data store, or chromosome [...] determines your genetic identity. [Gross 1999, pp. 3-5]

Drexler's next conclusion, that the bio-to-nano analogy allows nano design to proceed quickly and progressively because the principles of macroscale design can simply be translated down, has met more resistance. Yet, the practice of nanotechnology shows that many in the field have accepted this point. Nanotechnology journals are filled with news about the latest nanogears, nanomotors, nanotrains, nanoabacuses, nanoshovels, and other macroscale machines and devices replicated on the nanoscale. The epistemic frame of nanotechnology relies heavily on 'simulations' of all sorts – not just mathematical models, but physical, miniaturized 'models' of macroscale artifacts. Often, these simulations take Drexler's translation from biological to mechanical at face value; for instance, in one well-known experiment (Soong *et al.* 2000), researchers bonded an adenosine triphosphate 'motor' protein to a substrate and used it to spin a small metal bar – an ATP 'engine' much like what Drexler describes. These physical simulations 'prove' new processes or techniques, yield components that can eventually be added together to form complex systems, and signpost nano's travel down a mechanically evolutionary, more or less Drexlerian, pathway. As George Whitesides describes this experiment, "at the very least, such research stimulates efforts to fabricate functional nanostructures by demonstrating that such structures can exist" (Whitesides & Love 2001).

Even Drexler's critics (such as Whitesides) often accede to this part of his thesis while pointing out that biology may offer lessons unknown to macroscale engineers – as in this recommendation by a prominent science editor and analyst:

> Why copy nature? Biomimetics has become such a popular buzzword that there is a risk of it becoming its own justification [...] Yet there is little in the history of chemistry, materials science or engineering to show that this need be so. The steam engine, internal combustion engine, jet engine, and rocket engine owe no debt to inspiration from nature [...] [Meanwhile m]icroelectronics continues its incredible shrinking act with only the barest hint of any weakening of Gordon Moore's 'law' [...] This reduction in scale brings engineering down to length scales comparable with the dimensions of cells or subcellular constituents. There are two ways in which one could respond to this situation. One could regard the coincidence in scale as irrelevant, since engineering's traditional methods and materials have nothing in common with those of the cell [...] The other option is to realize that the cell faces many, if not most, of the same challenges as we do [...] The ideal position lies, as ever, somewhere in between. I feel that the literal down-sizing of mechanical engineering popularized by nanotechnologists such as Eric Drexler – whereby every nanoscale device is fabricated from hard moving parts, cogs, bearings, pistons and camshafts – fails to acknowledge that there may be better, more inventive ways of engineering at this scale [...] On the other hand, we should remember that the cell's objectives are not necessarily the engineer's. [Ball 2002, pp. 13-16]

Note how this author, like Drexler, references everything about nano to an instructive past and a future shaped by rules such as Moore's Law. Note, too, though, how the author uses law-like observations about the evolution of science and engineering in the past to define a particular purview for nanotechnology now and in the future. Interestingly, though they share the use of this trope, Drexler and his critics disagree about how to apply it in defining the field. Drexler sees the history and practice of engineering as providing *analogical* design cues for how to build things with atoms once we have mastered their precise control, and as giving a systems perspective that allows us to make enormous complexes of nanoscale machines work in coordinated ways – so-called 'nanofactories' that work almost exactly like macroscale factories, with conveyor belts and assembly lines and computer control. Yet, for Drexler there is little or no *genealogical* connection between traditional engineering's march of miniaturization (the so-called 'top-down' approach) and molecular nanotechnology's atomic precision (the 'bottom-up' approach).

Non-Drexlerians, and some Drexler associates, though, describe engineering's unstoppable march *down* in length scale as converging with chemistry's and molecular biology's journey *upward* in the size of the entities they can comprehend. This convergence gives nano its character, and makes a unified study of the nanoscale a necessity. As Heini Rohrer, Nobel Prize-winning co-inventor of the scanning tunneling microscope, puts it,

> While solid-state science and technology have moved down from the millimeter to the nanometer scale, chemistry has simultaneously and independently progressed from the level of small, few-atom molecules to macromolecules of biological size [...] The nanometer age can thus be considered as a continuation of an ongoing development: for example, miniaturization in solid-state technology [and] increasing complexity in chemistry. [Rohrer 1995, p. 3]

Compare this with a very similar passage from a prominent Foresight Institute participant:

> In the years that followed [Feynman's 1959 talk], chemists and biologists focused on untangling the molecular structures that constitute materiality from the 'bottom up,' while physicists and electrical engineers devoted their efforts to building ever smaller machines from the 'top down' [...] The recent confluence of these two monumental efforts has produced an epochal cross-fertilization of knowledge – and the inevitable conceptual turbulence of two colliding world views [...] Nanotechnology arises out of this confluence and aims at building complex, atomically precise machines by the trillions. [Crandall 1999, p. 21]

Rohrer, Crandall, and others who write in this vein almost always include charts and graphs that correlate the two key variables of nanotechnological determinism: length scale and time. Rohrer, for example, includes a diagram with length on one axis and year on the other showing two converging lines: one for steadily *decreasing* size of the smallest structures that can be included in the 'made world' of engineering (microelectromechanical systems, semiconductor chip features, *etc.*); and the other for steadily *increasing* size of the largest molecules that make up part of the made world of chemistry (dendrimers, nanotubes, buckyballs, and so on).

Many writers frame nanotechnology with a chart describing conspicuous features and characteristic entities of length scales from the humanly familiar (usually one meter or centimeter – represented by a familiar animal such as a bee or a cat) to the sub-nanoscopic (one angstrom—represented by a hydrogen atom) and everything in between. Often, these writers juxtapose the chart of length scales with a chart of significant nanotechnological achievements and their dates; usually, such events include the birth dates of the more artificial epistemic materials in the length scale chart (*e.g.* buckyballs or integrated circuits), as well as the dates of invention of new ways to handle or characterize these materials (*e.g.* the electron or scanning tunneling microscopes). Almost always, though, this timeline includes exquisite outliers that make the history of nanotechnology unfathomably deep; for instance, the first two items in a nano-timeline from *Scientific American* are "*3.5 billion years ago* the first living cells emerge" and "*400 B.C.* Democritus coins the word 'atom'" (Stix 2001, p. 36). This is one of the most pervasive and interesting characteristics of nanotechnology, common to Drexlerians and non-Drexlerians alike. Drexler and his allies tend to focus on the very ancient *biological* precursors of nanotechnology, since this helps them make the analogy between biological and artificial nanomachines, and because Drexler has worked hard to limit the scope of 'nanotechnology' to only those activities that involve precise positioning of individual atoms. This is a more limited scope with fewer precursors in human history than that offered under, for example, the National Nanotechnology Initiative's definition of the field. Those outside the Drexler camp, meanwhile, are more likely to point out very old *craft* activities that would today count as 'nanotechnology':

> The process of nanofabrication, in particular the making of gold nanodots, is not new. Much of the color in the stained glass windows found in medieval and Victorian churches and some of the glazes found in ancient pottery depend on the fact that nanoscale properties of materials are different from macroscale properties [...] In some senses, the first nanotechnologists were actually glass workers in medieval forges rather than the bunny-suited workers in a modern semiconductor plant. Clearly the glaziers did not understand why what they did to gold produced the colors it did, but we do now. [Ratner & Ratner 2003, pp. 13-14]

The last part of this quote shows some of the epistemic consequences of nanotechnology's non-presentism. Nano, in this formulation, produces new *knowledge* that maps onto old *practice*. What makes nano new is that it brings *understanding* where before there was only *doing*. Though nanodots in stained glass are an extreme example, the epistemic shyness of nano, and its strong predilection for creating knowledge by creating *nano-things*, does encourage nanoists to mine past work for present results. Indeed, in one of nano's most important constituent communities, surface science, researchers are exploring practices that in the past they rejected specifically because they yielded non-epistemic materials.

> [Surface scientists] were interested in understanding the science base of what was necessary in order to grow materials of interest to the electronics community [...] You had to understand the surface in a lot of detail, how you grew the thin film on top of it and kept a very fine, smooth surface. A tremendous amount of work had to go into the preparation of the surface, understanding how things settled down, what structures were there, how you varied the process and conditions to get it. One of the amusing things to me was that for many decades the people who were trying to grow these superlattices worked very hard to get

> these perfectly smooth surfaces, which they needed. So anytime they found conditions in which you got a non-flat surface, they would turn around and run the other direction. Which was appropriate at the time. Now when we get into the nano, what they've discovered is that some of those things they were trying desperately to avoid back then were giving them 'ordered nanostructures'. Which was killing them at the time, but now becomes of a high degree of interest [...] Some of the things that were the poison back then now become the candy that you can go back and say 'ooh, yeah!' We turned and ran the other direction back then, but let's go back and try 'what happens if we push harder, can we now enhance that growth rate and give us these little pyramidal islands?'

I can only make exploratory gestures toward a better understanding of nano's orientation to the past here, but it seems so unusual and so central to the current framing of nanotechnology that it deserves more intensive study. It is possible that nano shares this kind of rhetoric with other non-presentist fields like astronomy, where participants orient explicitly to pre-scientific ancestors of the modern discipline (and even occasionally use the work of those ancestors to better understand the history of the objects of study).

It is also possible that these kinds of statements are necessary *now*, when nanotechnology is being defined and woven into a coherent discipline. For instance, rhetoric of this sort certainly helps nano proponents convince various publics that nano has a long and hence non-threatening lineage. This is similar to attempts by biotechnology companies to persuade the public that genetic engineering is simply the latest variant of an ancient tradition of plant breeding, animal husbandry, and beer-making, rather than the dawn of a scary new Frankenstein-era. The need for boundary-drawing and credence also seems to be at the root of nanoists' constant search for prominent researchers of the past who can be recast as heroes of proto-nanotechnology. This is especially true of Richard Feynman, whose obscure after-dinner speech from the 1959 American Physical Society meeting, 'There's Plenty of Room at the Bottom' (Feynman 1999), has been taken up as a herald of all aspects of the new field. The phenomenon is by no means limited to Feynman, though – icons like Einstein, Schrödinger, and von Neumann are also routinely invoked as having done nano before there was nano. Nanoists carry their boundary-drawing struggles to the past in other ways as well. It is difficult, for instance, to find a description of nanotechnology that does not call it 'the next' X or Y. Even the official slogan of the National Nanotechnology Initiative is that nano is the "second industrial revolution" (Anonymous 2002, p. 3). Different participants cast around for different historical models and different kinds of lessons to draw from them. Drexler, for one, usually points to fields – such as space travel, computing, or aviation – with individual, visionary founders (Goddard, Babbage, da Vinci) who were unsuccessful in their own time but eventually proven correct. For investors, or those trying to attract capital, the relevant examples are the rise of the biotech industry, the dot-com boom and bust, or the law-like progress of semiconductor manufacturing. Finally, those who are trying to build national infrastructures *for* nanotechnology, or who are trying to make nano part *of* the global economy, often draw analogies to the giant technological systems of the past.

There is a curious, though surely quite common, mixing of technological and social determinism in this way of arguing. On the one hand, it is clear that nano is not completely determined on its own merits; societies have some choice in molding it to look more like some historical models than others. Yet, proponents and critics both seem to say that once we figure out whether nano looks more like the computer industry or the electricity industry or the biotech industry then we can predict how it will proceed. Societies have some choice at the highest level (do we do nano at all?), but once they dip their toes in the water they will be swept along; and if they do not jump in the river now, their competitors will quickly outdistance them. Take, for instance, this assertion from a supporter of the US "21st Century Nanotechnology Research and Development Act":

> From the dawn of modern agriculture to aerospace to the launching of the Information Age, government support has been a powerful catalyst to drive basic research and accelerate technology from the laboratory to the marketplace. In industry after industry, one sees the same pattern: federal dollars encourage early discoveries in a new technology, which then attracts private investment, which then grows into a successful industry, with large employers and many jobs [...] We are now at a critical juncture in our technological evolution, and timely passage of this bill will go far to assuring American leadership in the global economy [...] We see other governments of the European Union and East Asian nations investing heavily in major nanotechnology research and development centers. The hard reality is that the worldwide race for preeminence in nanotechnology is on, and America must push to stay in the lead. [Swami 2002]

Indeed, this is exactly the sort of reasoning Drexler uses to motivate the founding of the Foresight Institute and his continuing efforts to describe the inevitably coming, but still able-to-be-influenced, nano-future:

> Some force in the world (whether trustworthy or not) will take the lead in developing assemblers; call it the 'leading force.' Because of the strategic importance of assemblers, the leading force will presumably be some organization or institution that is effectively controlled by some government or group of governments [...] Design-ahead can help the leading force prepare, yet even vigorous, foresighted action seems inadequate to prevent a time of danger. [Drexler 1990, p. 182]

Drexler and his critics agree, then, that nano is on its way whether we choose to be part of it or not. They agree, too, that when it arrives, everything will be different; society will have to adapt to nano much more than the other way around. Drexler's vision of the post-nano world is perhaps the more sweeping, and it has clearly influenced the vivid, exquisitely imaginative depictions of science fiction writers such as Neal Stephenson and Kathleen Ann Goonan (Milburn 2002). Interestingly, though, Drexler originally wrote in *Engines of Creation* that a post-nano future would leave us *free* from technological determinism; we would inhabit a world made so radically malleable by nano that we could be liberated from the constraints of any one technological system:

> [The modern technological] system now sprawls across continents, entangling people in a global web. It has offered escape from the toil of subsistence farming, lengthening lives and bringing wealth, but at a cost that some consider too high. Nanotechnology will open new choices. Self-replicating systems will be able to provide food, health care, shelter, and other necessities. They will accomplish this without bureaucracies or large factories. Small, self-sufficient communities can reap the benefits. One test of the freedom a technology offers is whether it frees people to return to primitive ways of life. Modern technology fails this test; molecular technology succeeds. As a test, imagine returning to a stone-age style of life—not by simply ignoring molecular technology, but while using it. [Drexler 1990, p. 235]

As Stefan Helmreich (1998) has pointed out, this theme of radical liberation made possible by new technologies is common in futurist circles: whether freedom from the earth (space travel), from the body (artificial intelligence and artificial life), or from death (Drexler's most-cherished application of nano is to allow frozen corpses to be reanimated and healed, allowing immortality for anyone born today). The freedom enabled by the massive changes brought on by nano is not particular to Drexler alone, though. For instance, some of his staunchest critics among practicing nanotechnologists and policy makers promote the idea that nano is the key to a transhumanist future, in which the very definition of human capabilities will have to be redefined. Even a die-hard Drexler-skeptic like George Whitesides sees a nano-future that bears little resemblance to today:

> [N]anoscale machines already do exist, in the form of the functional molecular components of living cells [...] What are the most interesting designs to use for future nanomachines? And what, if any, risks would they pose? [...] [A]s for ravaging the earth: in a sense, collections of biological cells already have ravaged the earth. Before life emerged, the planet was very different from the way it is today. Its surface was made of inorganic minerals; its atmosphere was rich in carbon dioxide. Life rapidly and completely remodeled the planet: it contaminated the pristine surface with microorganisms, plants and organic materials derived from them; it largely removed the carbon dioxide from the atmosphere and injected enormous quantities of oxygen. Overall, a radical

> change. Cells – self-replicating collections of molecular nanomachines – completely transformed the surface and the atmosphere of our planet. We do not normally think of this transformation as 'ravaging the planet,' because we thrive in the present conditions, but an outside observer might have thought otherwise. So the issue is not whether nanoscale machines can exist – they already do – or whether they can be important – we often consider ourselves as demonstrations that they are – but rather where we should look for new ideas for design. [Whitesides 2001, pp. 78-79]

This quote from Whitesides sums up all three of the arguments used by nanoists of all stripes that fall well within classic notions of technological determinism: that nano is inevitable; that it will develop with its own progressive, internal logic (though we have some choice whether to follow the logic of biology or engineering); and that nano *itself*, beyond the control of society, will completely transform the world. Indeed, with regard to the latter, Whitesides plays with fears of the so-called 'grey goo' problem – a catastrophic scenario in which nanomachines become so completely autonomous and uninfluenced by social considerations that they run amok and destroy life as we know it (perhaps the most extreme form of technological determinism imaginable). Whitesides also displays some of the peculiarities in the way nanoists handle determinist arguments, particularly in his consistent non-presentism – it is difficult to imagine other sciences where events of billions of years ago would so consistently be invoked unless those events were themselves the objects of study (as is the case in geology or cosmology but not in nanotechnology). I conclude by examining two more tropes that nanoists have applied as technologically determinist arguments, but that they have applied in such unusual ways that they tell us a great deal about the field's epistemic and practical frame. The first, which has been discussed much more thoroughly elsewhere by Alfred Nordmann (2004), might be called the trope of manifest destiny. Nordmann points out that much of the epistemic shyness of nano research comes from practitioners' conceptualization of the field as focused on a space (the nanoscale) rather than a characteristic set of materials or practices or concepts. Nano is oriented much more to expanding human *control* over larger areas of the nanoscale and the entities that inhabit it than to learning anything fundamental about 'nature' or 'reality'.

As we have seen, control over the nanoscale has long been an aim of some of nanotechnology's constituent communities, such as chemistry or surface science; but in those disciplines *control* was seen as a means to generating fundamental knowledge about a few characteristic materials (*i.e.*, about creating an epistemically amenable 'made world'), rather than (as in nanotechnology) as an end unto itself.

Nanoists often represent their relation to this new place, the nanoscale, as one of dominance and entitlement—it is their manifest destiny to explore, control, and remake this undiscovered country. Roots for this trope can clearly be found in Drexler's original formulation of the field; after all, the futurist tradition, particularly with regard to space travel, has long been obsessed with creating new 'final frontiers' where technological achievement necessitates the outward expansion of control and exploration. Nano, at least in the United States, is merely the latest effort to engage what David Nye has called the 'American technological sublime' (Nye 1994)—the attempt, so central to America's self-conception, to create something transcendent and beyond humanity *through* artificial structures. Drexler's early work radiates the technological sublime, with his talk of immortality, space travel, and radical transhumanism made possible by molecular assemblers. Moreover, his description of the imminent development of the nanoscale closely resembles a narrative of American frontier expansion: from the first sighting of land (the imaging of atoms with a scanning tunneling microscope), to interactions with 'natives' (biological nanomachines), to the appropriation of some technologies from those natives and the wholesale importation of simple non-native technologies (nanoscale bearings, gears, *etc.*), and finally the imposition of state control over the lawless nanoscale and widespread industrialization through the proliferation of nano-factories.

Non-Drexlerians, too, see just as certain a manifest nanodestiny. After all, the US National Nanotechnology Initiative calls its founding document 'Small Wonders, Endless Frontiers' (Anonymous 2002) – a combination of the technological sublime, frontier expansion into the nanoscale, and a postwar American tradition, going back to Vannevar Bush's (1945) *Science, the Endless Frontier*, of seeing science as the next arena for the nation's manifest destiny. Nanoists perform this destiny in a variety of ways in their research practices. For instance, in coming of age at the same time as widespread computing, nanotechnology has made much more extensive use of computer graphics than any traditional discipline. When they can, nanoists use this software to render images of their made world as breathtaking landscapes of wide-open vistas, often portrayed in the coloring of the deserts of the American West. Often, such images possess a great deal of visual *éclat*, but are more difficult to integrate with theory than more traditional, non-perspectival representations. At the same time, nanoists often stake a claim to these landscapes by literally writing their ownership right into the material itself – through various nanolithography techniques they can, and do, inscribe their names, their favorite phrases, and, inevitably, a series of flags, maps, and patriotic proclamations. Again, this goes to the epistemic heart of nanotechnology – it is a field where 'proof' can be achieved just as readily by writing one's name as by more traditional methods for assuring the rigor of knowledge. It is necessary only to show that one *owns* a patch of the nanoscale to have contributed to nano's body of knowledge.

The second, related, trope stems from nanoists' predilection for what I have called elsewhere 'nanopresence' (Mody 2004). Nanopresence is, basically, the endowment of nano-objects with familiarity, tangibility, and even personality – the creation of a sense that they can be touched, that they are ordinary and quotidian objects of interaction. As the name implies, nanopresence owes some debt to Heidegger's thoughts on the nature of technology and his distinction between ready-to-hand and present-at-hand (Heidegger 1962). In Heidegger's formulation, technological artifacts have two quite distinct phenomenological casts – one we experience when we regard the artifact as an object, something that can be theorized about, that can be thought about apart from the act of actually using it; the other is the artifact as we experience it when we are using it, when we and the tool become extensions of each other and we cannot pause to consider the tool apart from how we actively engage with it.

Nanotechnology can, in many respects, be seen as the coordinated attempt to recast nanoscale objects as ready-to-hand tools, to move past the theories and epistemic pretensions of nano's constituent communities and instead use their knowledge to actively engage with the nanoscale. Interestingly, 'handedness' has a very long history in nanotechnology. In Richard Feynman's original 'There's Plenty of Room at the Bottom' speech, he lays out a vision of miniaturization in which he imagines a linked chain of progressively smaller 'hands' that allow us to make progressively tinier bits of the world 'ready-to-hand'.

> How do we make such a tiny mechanism? [...] [I]n the atomic energy plants they have materials and machines that they can't handle directly because they have become radioactive. To unscrew nuts and bolts and so on, they have a set of master and slave hands, so that by operating a set of levers here, you control the 'hands' there, and can turn them this way and that so you can handle things quite nicely [...] Now, I want to build much the same device – a master-slave system which operates electrically. But I want the slaves to be made especially carefully by modern large-scale machinists so that they are one-fourth the scale of the 'hands' that you ordinarily maneuver. So you have a scheme by which you can do things at one-quarter scale anyway [¼] Aha! So I manufacture a quarter-size lathe; I manufacture quarter-size tools; and I make, at one-quarter scale, still another set of hands again relatively one-quarter size! [...] Well, you get the principle from there on. [Feynman 1999]

As Colin Milburn and Ed Regis point out, Feynman probably got this idea from a short story by Robert Heinlein. This is not unusual for the field; indeed, it is one of the oddities of nano that it relies so much on science fiction to supply thought experiments and fodder for 'proofs of concept'. It is perhaps not surprising, though, that nano, with its predilection for simulation and the re-enchantment of the material world, should recognize an affinity with fiction, the art of making the unreal seem experienced and ready-to-hand.

Social constructionists have critiqued Heidegger's formulation as containing its own kind of technological determinism – the tool that is ready-to-hand seems pinned to one and only one use, whereas with most technologies users show a great deal of flexibility in alternately regarding and using artifacts in idiosyncratic ways. Analysts interested in exploring this issue and pushing the Heideggerian interpretation toward a more nuanced position will find exquisite material in nanotechnology. On the one hand, nanoists have really embraced the handedness of Feynman's original vision. For instance, almost incontrovertibly the most famous nano image thus far produced is Don Eigler's (Eigler & Schweizer 1990) 'IBM' written with individual xenon atoms positioned by a scanning tunneling microscope (STM). Eigler has his STM set up such that one can simply move the STM tip around with a mouse, click on an atom, drag it to where it should go, and release it. It is almost impossible when doing so to think of the atom as an object of theory, as the heuristic fiction so beloved of positivists a century ago. Instead, mouse and atom are simply ready-to-hand, ready to be moved around, placed into various two-dimensional structures, and generally experienced as a bright spot on a computer screen with which one has some haptic engagement.

Other nanoists take this several steps further. Among nano experimentalists who specialize in building very high-end instrumentation (particularly in the scanning tunneling and atomic force microscopy community) there has been a rush in the past few years to incorporate more and more sensory engagement into their instruments, to make the nanoscale ever more ready-to-hand. Builders of molecule pullers, such as Paul Hansma (Viani *et al.* 1999) and Hermann Gaub (Clausen-Schaumann *et al.* 2000), for instance, have designed instruments that slowly pry apart the internal domains of complex biomolecules. Some of these pullers have built-in resistance on the controls – the operator can actually 'feel' the domains popping, rather like feeling the jerks of a fish caught on the end of a line. Other pullers have a simple circuit that allows the shaking of the puller cantilever to be translated into a sound; operators can *listen* to the molecular domains popping. One puller designer describes how these instruments provoke a feeling that the nanoscale is ready-to-hand, and how this handedness is epistemically (and commercially) useful:

> It's really good at [trade] shows too, because if you're actually introducing a subject to somebody, thermal noise for example, it's one thing to explain it to them, it's another to hand them a pair of headphones and say 'look, this is what thermal noise is' and you can explain the concepts of damping and things like that and how the spectrum shifts because it's totally obvious when you just hear it, it's like 'yeah of course, that's what's happening.'

Perhaps the most well-known attempt in this direction is the Nanomanipulator at the University of North Carolina (Guthold *et al.* 2000). There, Rich Superfine's group has built an atomic force microscope with special haptic feedbacks and virtual reality controls. Users can 'stand' in the landscape of the nanoscale, they can 'feel' how rough or smooth nanoscopic surfaces are, and they can even nudge nano-objects (such as buckytubes) around.

At the same time, nanoists enjoy playing with the handedness of the nano realm by pushing their audience into an ambiguous state where images and representations oscillate between the ready-to-hand and the present-at-hand. Witness all the nano-plows and nano-shovels and nano-trains and abacuses and whatnot – all these nano-artifacts seem like tailor-made tools in Heidegger's simple, ready-to-hand kit. Again, this plays well to nano's epistemic shyness; just seeing an image of nanoscale abacus or guitar or train and apprehending these objects instantly *as such* makes the audience's first experience of them an engaged, ready-to-hand involvement rather than distanced, theoretical or conceptual observation. Yet, that instant recognition carries with it a simultaneous wonder and shock—the nano-object is all too familiar, yet all too different and exotic. The nanoscale has become a place that tourists can visit, where everything is different, yet exactly the same—all the building blocks are atoms, at which we should wonder, but they are being used to make ordinary, familiar, everyday objects whose use is something we intuit rather than theorize about. For now, I have to turn my spade in digging at this phenomenon—I am not sure how to read the handedness of nano, though it seems clear many layers of practice and rhetoric are involved. What I would encourage as this, hopefully, becomes a topic for analysis is that we remember that nanoists' tweaking of intuitive understandings is done, usually, in a spirit of fun and play.

From Feynman's first playful call for researchers to make tiny motors and write words on the head of a pin to today's silicon zoo of tiny guitars, flags, signatures, and so forth, nanoists have let themselves be seen to be having fun. The debates between Drexler and his critics have taken an acrid and unpleasant tone in the past few years, but analysts of nano should not take this to be the whole show. For many practitioners, nano is still a bit of a put-on, a bandwagon whose content they do not quite understand but which they are trying to make the best of. This 'making do' has a distinctively light-hearted cast, as practitioners trot out parlor tricks that double as proofs of concept, and as they avoid interdisciplinary frictions by sticking to relatively uncontroversial play. Nanoists have created a technological sublime, but in shrinking the dimensions of the sublime to such an extent, they have made it provoke both awe and a bit of laughter.

More generally, we should keep this playfulness in mind in examining what uses nanoists make of determinist arguments. For many nanoists, nano *is* inevitable and (nano)technology does drive (some of) history. Yet there is little fatalism in the nano community; practitioners seem more eager to ride the tiger of nano than they are apprehensive that they will be crushed by it. Nanoists seem, for instance, willing to *play* with the design logic made possible by the analogy between biological and artificial nanomachines. While they agree that everything will change because of the new technology, nanoists have used this agreement to inspire both serious discussion of *how* to prepare, as well as dramatic, sometimes inspiring, flights of fancy about *what* to prepare for. Nano is still an incoherent mass of often conflicting communities. Determinist arguments advance the particular interests of various kinds of practitioners within this mass, as well as various critics and supporters on the outside. If we are to understand nano, we must see how participants build these arguments into their practices, and how they do so in ways that allow them to live with the field's current incoherence.

REFERENCES

Beck, J.S., J.C. Vartuli, W.J. Roth, M.E. Leonowicz, C.T. Kresge, K.D. Schmitt, C.T.-W. Chu, D.H. Olsen, E.W. Shepard, S.B. McCullen, J.B. Higgins, and J.L. Schlenker. 1992. *J. Am. Chem. Soc*. 114:10834.

Bensaude-Vincent, B.: 2004, 'Two Cultures of Nanotechnology?', *Hyle: International Journal for Philosophy of Chemistry*, 10(2), 65-82.

Bowes, C.L., A. Malek, and G.A. Ozin. 1996. *Chem. Vap. Deposition* 2:97.

Bueno, O.: 2004, 'Von Neumann, Self-Reproduction and the Constitution of Nanophenomena', in: D. Baird, A. Nordmann & J. Schummer (eds.), *Discovering the Nanoscale*, Amsterdam: IOS Press, pp. 101-115.

Chen, C-Y., S.L. Burkett, H.-X. Li, and M.E. Davis. 1993. *Microporous Mater*. 2:27.

Dresselhaus, M.S., G. Dresselhaus, and P. Eklund. 1996. *Science of fullerenes and carbon nanotubes*. San Diego: Academic Press.

Dupuy, J.-P.: 2004, 'Complexity and Uncertainty: A Prudential Approach To Nanotechnology', in: European Commission (Community Health and Consumer Protection): *Nanotechnologies: A Preliminary Risk Analysis on the Basis of a Workshop, Brussels, 1-2 March 2004*, pp. 71-93 (www.europa.eu.int/comm/health/ ph_risk/documents/ev_20040301_en.pdf).

Einsiedel, E.F.; Goldenberg, L.: 2004, 'Dwarfing the Social? Nanotechnology Lessons from the Biotechnology Front', *Bulletin of Science, Technology & Society*, 24, 28-33.

ETC: 2003b, 'No Small Matter II: The Case for a Global Moratorium ' Size Matters!' *Occasional Paper Series* 7(1) [www.etcgroup.org/documents/Occ.Paper_Nanosafety.pdf].

European Comission: *European Workshop on Social and Economic Research on Nanotechnologies and Nanosciences, Brussels; 14-15 April 2004* [www.stage-research.net/STAGE/PAGES/Nano.html].

F. Buot, "Mesoscopic Physics and Nanoelectronics: Nanoscience and Nanotechnology," *Physics Reports*, pp.73-174, 1993.

Feynman, R., "There's Plenty of Room at the Bottom: An invitation to Enter a New Field of Physics," Talk at the Annual Meeting of the American Physical Society, 29 December 1959. Reprinted in Appendix B of Crandall and Lewis. See citation above.

Fleischer, Th.; Decker, M. & Fiedeler, U. (eds.): 2004, *Große Aufmerksamkeit für kleine Welten ' Nanotechnologie und ihre Folgen*, special issue of *Technikfolgenabschätzung ' Theorie und Praxis*, 13 (2), 5-85 [www.itas.fzk.de/tatup/042/inhalt.htm].

Fogelberg, H. & Glimell, H.: 2003, *Bringing Visibility to the Invisible: Towards A Social Understanding of Nanotechnology*, Göteborg: Göteborg University [www.sts.gu.se/publications/STS_report_6.pdf].

Fogelberg, H.: 2003, 'The Material Culture of Nanotechnology', in: H. Fogelberg & H. Glimell, *Bringing Visibility to the Invisible: Towards A Social Understanding of Nanotechnology*, Göteborg: Göteborg University, pp. 99-114.

Glimell, H.: 2001, 'Challenging Limits ' Excerpts from an Emerging Ethnography of Nano Physicists', in: H. Glimell & O. Johlin (eds.), *The Social Production of Technology: On the Everyday Life with Things*, Göteburg, SE: BAS Publisher, chapter 7, pp. 111-131 (reprinted in: H. Fogelberg & H. Glimell, *Bringing Visibility to the Invisible: Towards A Social Understanding of Nanotechnology*, Göteborg: Göteborg University, pp. 115-139.

Glimell, H.: 2003, 'A Nano Narrative: The Mircopolitics of a New Generic Enabling Technology', in: H. Fogelberg & H. Glimell, *Bringing Visibility to the Invisible: Towards A Social Understanding of Nanotechnology*, Göteborg: Göteborg University, pp. 55-77 [www.sts.gu.se/publications/STS_report_6.pdf].

Glimell, H.: 2004, 'Grand Visions and Lilliput Politics: Staging the Exploration of the 'Endless Frontier", in: D. Baird, A. Nordmann & J. Schummer (eds.), *Discovering the Nanoscale*, Amsterdam: IOS Press, pp. 231-246.

Gorman, M.E.; Groves, J.F. & Shrager, J.: 2004, 'Societal Dimensions of Nanotechnology as a Trading Zone: Results from a Pilot Project', in: D. Baird, A. Nordmann & J. Schummer (eds.), *Discovering the Nanoscale*, Amsterdam: IOS Press, pp. 63-73.

Grinbaum, A.: 2004, 'La condition de l'homme moderne et les nanotechnologies', in: G. Nivat (ed.), *Les limites de l´humain. 39èmes Rencontres Internationales de Genève*, L´Age d´Homme, Genève, p. 141.

Gupta, V.K. & Pangannaya, N.B.: 2000, 'Carbon nanotubes: bibliometric analysis of patents', *World Patent Information*, 22, 185-189.

Haruta, M. 1997. *Catalyst surveys of Japan* 1:61 and references therein.

Hessenbruch, A.: 2004, 'Nanotechnology and the Negotiation of Novelty', in: D. Baird, A. Nordmann & J. Schummer (eds.), *Discovering the Nanoscale*, Amsterdam: IOS Press, pp. 135-144.

Huo, Q., D.I. Margolese, U. Ciesla, P. Feng, T.E. Gier, P. Sieger, R. Leon, P.M. Petroff, F. Schuth, and G.D. Stucky. 1994. *Nature* 368:317.

Johansson, M.: 2003, 'Plenty of room at the bottom: Towards an anthropology of nanoscience', *Anthropology Today*, 19 (No. 6), 3-6.

Jounet, C., W.K. Maser, P. Bernier, A. Loiseau, M. Lamy de la Chapelle, S. Lefrant, P. Deniard, R. Lee, and J.E. Fischer. 1997. *Nature* 388:756.

Khushf, G.: 2004, 'A Hierarchical Architecture for Nano-scale Science and Technology: Taking Stock of the Claims About Science Made By Advocates of NBIC Convergence', in: D. Baird, A. Nordmann & J. Schummer (eds.), *Discovering the Nanoscale*, Amsterdam: IOS Press, pp. 21-33.

Krätschmer, W., L.D. Lamb, K. Fostiropoulos, and D.R. Huffman. 1990. *Nature* 347:354.

Kupperman, A., S. Nadimi, S. Oliver, G. Ozin, J. Garcés, and M. Olken. 1993. *Nature* 365:239.

Kuusi, O.; Meyer, M.: 2002, 'Technological generalizations and leitbilder ' the anticipation of technological opportunities', *Technological Forecasting & Social Change*, 69, 625-639.

Landon, B.: 2004, 'Less is More: Much Less is Much More: The Insistent Allure of Nanotechnology Narratives in Science Fiction', in: N.K. Hayles (ed.), *Nanoculture: Implications of the New Technoscience*, Bristol, UK: Intellect Books, pp. 131-146.

Laszlo, P.: 2004, 'Is There Life After Partington?', *Hyle: International Journal for Philosophy of Chemistry*, 10(2), 169-178.

Lenhard, J.: 2004, 'Nanoscience and the Janus-Faced Character of Simulations', in: D. Baird, A. Nordmann & J. Schummer (eds.), *Discovering the Nanoscale*, Amsterdam: IOS Press, pp. 93-100.

López, J.: 2004, 'Bridging the Gaps: Science Fiction in Nanotechnology', *Hyle: International Journal for Philosophy of Chemistry*, 10(2), 129-152.

Lösch, A.: 2004, 'Nanomedicine and Space: Discursive Orders of Mediating Innovations', in: D. Baird, A. Nordmann & J. Schummer (eds.), *Discovering the Nanoscale*, Amsterdam: IOS Press, pp. 193-202.

Marshall, K.: 2004, 'Atomizing the Risk Technology', in: N.K. Hayles (ed.), *Nanoculture: Implications of the New Technoscience*, Bristol, UK: Intellect Books, pp. 147-160.

Mehta, M.D.: 2002, 'Nanoscience and Nanotechnology: Assessing the Nature of Innovation in These Fields', *Bulletin of Science, Technology & Society*, 22(4), 269-273.

Mehta, M.D.: 2004, 'From Biotechnology to Nanotechnology: What Can We Learn from Earlier Technologies?, *Bulletin of Science, Technology & Society*, 24, 34-39.

Meyer, M. & Kuusi, O.: 2004, 'Nanotechnology: Generalizations in an Interdisciplinary Field of Science and Technology', *Hyle: International Journal for Philosophy of Chemistry*, 10(2), 153-168.

Meyer, M.: 2000a, 'Does science push technology? Patents citing scientific literature', *Research Policy*, 29, 409-434.

Meyer, M.: 2000b, 'Patent citations in a novel field of technology: What can they tell about interactions of emerging communities of science and technology?', *Scientometrics*, 48, 151-178.

Meyer, M.: 2001a, 'Patent citations in a novel field of technology: An exploration of nano-science and nano-technology', *Scientometrics*, 51, 163-183.

Meyer, M.: 2001b, 'Socio-economic Research on Nanoscale Science and Technology: A European Overview and Illustration', in: M.C. Roco and W.S. Bainbridge (eds.), *Societal Implications of Nanoscience and Nanotechnology*, Dordrecht: Kluwer, pp. 217-241.

Milburn, C.: 2002, 'Nanotechnology in the age of post-human engineering: science fiction as science', *Configurations*, 10, 261-295 [reprinted in: N.K. Hayles (ed.), *Nanoculture: Implications of the New Technoscience*, Bristol, UK: Intellect Books, 2004, pp. 109-130].

Milburn, C.: 2004, 'Nano/Splatter: Disintegrating the Postbiological Body', *New Literary History*, 35 (in print).

Mnyusiwalla, A.; Abdallah, S.D.; Singer, P.A.: 2003, 'Mind the gap: science and ethics in nanotechnology', *Nanotechnology*, 14, R9-R13.

Mody, C.C.M.: 2004, 'How Probe Microscopists Became Nanotechnologists', in: D. Baird, A. Nordmann & J. Schummer (eds.), *Discovering the Nanoscale*, Amsterdam: IOS Press, pp. 119-133.

Mody, C.C.M.: 2004, 'Instruments in Training: The Growth of American Probe Microscopy in the 1980s', in: D. Kaiser (ed.), *Pedagogy and the Practice of Science: Producing Physical Scientists, 1800-2000*, Cambridge, MA: MIT Press (forthcoming).

Mody, C.C.M.: 2004, 'Small, but Determined: Technological Determinism in Nanoscience', *Hyle: International Journal for Philosophy of Chemistry*, 10(2), 99-128.

Mody, C.C.M.: 2004, *Crafting the Tools of Knowledge: The Invention, Spread, and Commercialization of Probe Microscopy, 1960-2000*, Ph.D. dissertation, Cornell University.

Montemerlo, M.S., Love, J.C., Opiteck, G.J., Goldhaber, D. J., and Ellenbogen, J.C., "Technologies and Designs for Electronic Nanocomputers," MITRE Technical Report 96W0000044, The MITRE Corporation, McLean, VA, July 1996. For more information, send e-mail to nanotech@mitre.org.

Moor, J.H. & Weckert, J.: 2004, 'Nanoethics: Assessing the Nanoscale From an Ethical Point of View', in: D. Baird, A. Nordmann & J. Schummer (eds.), *Discovering the Nanoscale*, Amsterdam: IOS Press, pp. 301-310.

Nordmann, A.: 2004, 'Social Imagination for Nanotechnology', in: European Commission (Community Health and Consumer Protection): *Nanotechnologies: A Preliminary Risk Analysis on the Basis of a Workshop, Brussels, 1-2 March 2004*, pp. 111-113 [www.europa.eu.int/comm/health/ph_risk/documents/ev_20040301_en.pdf].

Nordmann, A.: 2004, 'Was ist TechnoWissenschaft? ' Zum Wandel der Wissenschaftskultur am Beispiel von Nanoforschung und Bionik', in: T. Rossmann & C. Tropea (eds.), *Bionik ' Neue Forschungsergebnisse aus Natur-, Ingenieur- und Geisteswissenschaften*, Berlin: Springer, 2004 (forthcoming)

Paschen, H.; Coenen, C.; Fleischer, T.; Grünwald, R.; Oertel, D. & Revermann, C.: 2004, *Nanotechnologie: Forschung, Entwicklung, Anwendung*, Berlin: Springer (*Nanotechnologie, TAB-Arbeitsbericht 92*, Berlin: Büro für Technikfolgen-Abschätzung beim Deutschen Bundestag).

Pitt, J.C.: 2004, 'The Epistemology of the Very Small', in: D. Baird, A. Nordmann & J. Schummer (eds.), *Discovering the Nanoscale*, Amsterdam: IOS Press, pp. 157-163.

Pressman, J.: 2004, 'Nano Narrative: A Parable from Electronic Literature', in: N.K. Hayles (ed.), *Nanoculture: Implications of the New Technoscience*, Bristol, UK: Intellect Books, pp. 191-200.

Prigogine, I., and S. Rice. 1988. *Advances in chemical physics*, Vol. 70, Parts 1 & 2. New York: J. Wiley.

R. T. Bate, "Nanoelectronics," *Nanotechnology*, Vol. 1, pp. 1-7, 1990.

Rademann, K., B. Kaiser, U. Even, F. Hensel. 1987. *Phys. Rev. Lett*. 59:2319.

Rao, C.N.R., B.C. Satishkumar, and A. Govindaraj. 1997. *Chem. Commun*. 1581.

Rao, M.B., and S. Sircar. 1993. *Gas Separation and Purification* 7:279.

Reed, M.A., "Quantum Dots," *Scientific American,* January 1993, pp. 118-123.

Roberts, J.A.: 2004, 'Deciding the Future of Nanotechnologies: Legal Perspectives on Issues of Democracy and Technology', in: D. Baird, A. Nordmann & J. Schummer (eds.), *Discovering the Nanoscale*, Amsterdam: IOS Press, pp. 247-255.

Robinson, C.: 2004, 'Images in NanoScience/Technology', in: D. Baird, A. Nordmann & J. Schummer (eds.), *Discovering the Nanoscale*, Amsterdam: IOS Press, pp. 165-169.

Robison, W.L.: 2004, 'Nano-Ethics', in: D. Baird, A. Nordmann & J. Schummer (eds.), *Discovering the Nanoscale*, Amsterdam: IOS Press, pp. 285-300.

Roco, M.C. & Bainbridge, W.S. (eds.): 2001, *Societal implications of nanoscience and nanotechnology*, (Proceedings of a workshop organized by the National Science Foundation, September 28-29, 2000), Kluwer: Dordrecht [available online at http://itri.loyola.edu/nano/societalimpact/nanosi.pdf]

Suchman, M.: 2002, 'Social Science and Nanotechnologies', M. Roco & R. Tomellini (eds.), *Nanotechnology ' Revolutionary Opportunities and Societal Implications*, Luxembourg: European Communities, pp. 95-99.

2

Nanotech Research and Society: An Overview

Nanotech Research, Nanoscience and Nanotubes: Applications & Methods

The term nanotube is normally used to refer to the carbon nanotube, which has received enormous attention from researchers over the last few years and promises, along with close relatives such as the nanohorn, a host of interesting applications. There are many other types of nanotube, from various inorganic kinds, such as those made from boron nitride, to organic ones, such as those made from self-assembling cyclic peptides (protein components) or from naturally-occurring heat shock proteins (extracted from bacteria that thrive in extreme environments). However, carbon nanotubes excite the most interest, promise the greatest variety of applications, and currently appear to have by far the highest commercial potential. Only carbon nanotubes will be covered in this chapter. Carbon nanotubes are often referred to in the press, including the scientific press, as if they were one consistent item. They are in fact a hugely varied range of structures, with similarly huge variations in properties and ease of production.

Adding to the confusion is the existence of long, thin, and often hollow, carbon fibers that have been called carbon nanotubes but have a quite different make-up from that of the nanotubes that scientists generally refer to. To distinguish these we will refer to them as carbon nanofibers. Carbon nanotubes were 'discovered' in 1991 by Sumio Iijima of NEC and are effectively long, thin cylinders of graphite, which you will be familiar with as the material in a pencil or as the basis of some lubricants. Graphite is made up of layers of carbon atoms arranged in a hexagonal lattice, like chicken wire. Though the chicken wire structure itself is very strong, the layers themselves are not chemically bonded to each other but held together by weak forces called Van der Waals. It is the sliding across each other of these layers that gives graphite its lubricating qualities and makes the mark on a piece of paper as you draw your pencil over it. Now imagine taking one of these sheets of chicken wire and rolling it up into a cylinder and joining the loose wire ends. The result is a tube that was once described by Richard Smalley (who shared the Nobel Prize for the discovery of a related form of carbon called buckminsterfullerene) as "in one direction . . . the strongest damn thing you'll ever make in the universe". In addition to their remarkable strength, which is usually quoted as 100 times that of steel at one-sixth of the weight (this is tensile strength—the ability to withstand a stretching force without breaking), carbon nanotubes have shown a surprising array of other properties. They can conduct heat as efficiently as most diamond (only diamond grown by deposition from a vapour is better), conduct electricity as efficiently as copper, yet also be semiconducting (like the materials that make up the chips in our computers). They can produce streams of electrons very efficiently (field emission), which can be used to create light in displays for televisions or computers, or even in domestic lighting, and they can enhance the fluorescence of materials they are close to. Their electrical properties can be made to change in the presence of certain substances or as a result of mechanical stress. Nanotubes within nanotubes can act like miniature springs and they can even be stuffed with other materials. Nanotubes and their variants hold promise for storing fuels such as hydrogen or methanol for use in fuel cells and they make good supports for catalysts. But let's look at some of the different types of nanotubes, and nanotube pretenders. One of the major classifications of carbon nanotubes is into single-walled varieties (SWNTs), which have a single cylindrical wall, and multi-walled varieties (MWNTs), which have cylinders within cylinders.

The lengths of both types vary greatly, depending on the way they are made, and are generally microscopic rather than nanoscopic, i.e. greater than 100 nanometers (a nanometer is a millionth of a millimeter). The aspect ratio (length divided by diameter) is typically greater than 100 and can be up to 10,000, but recently even this was made to look small. In May 2002 SWNT strands were made in which the SWNTs were claimed to be as long as 20 cm. Even more recently, the same group has made strands of SWNTs as long as 160 cm, but the precise make-up of these strands has not yet been made clear. A group in China has also found, purely by accident, that packs of relatively short carbon nanotubes can be drawn out into a bundle of fibers, making a thread only 0.2 millimeters in diameter but up to 30 centimeters long. The joins between the nanotubes in this thread represent a weakness but heating the thread has been found to increase the strength significantly, presumably through some sort of fusing of the individual tubes. These are the stars of the nanotube world, and somewhat reclusive ones at that, being much harder to make than the multi-walled variety. The oft-quoted amazing properties generally refer to SWNTs. As previously described, they are basically tubes of graphite and are normally capped at the ends, although the caps can be removed. The caps are made by mixing in some pentagons with the hexagons and are the reason that nanotubes are considered close cousins of buckminsterfullerene, a roughly spherical molecule made of sixty carbon atoms, that looks like a soccer ball and is named after the architect Buckminster Fuller (the word fullerene is used to refer to the variety of such molecular cages, some with more carbon atoms than buckminsterfullerene, and some with fewer). The theoretical minimum diameter of a carbon nanotube is around 0.4 nanometers, which is about as long as two silicon atoms side by side, and nanotubes this size have been made. Average diameters tend to be around the 1.2 nanometer mark, depending on the process used to create them. 8 SWNTs are more pliable than their multi-walled counterparts and can be twisted, flattened and bent into small circles or around sharp bends without breaking.

Discussions of the electrical behavior of carbon nanotubes usually relate to experiments on the single-walled variety. As we have said, they can be conducting, like metal (such nanotubes are often referred to as metallic nanotubes), or semiconducting, which means that the flow of current through them can be stepped up or down by varying an electrical field. The latter property has given rise to dreams of using nanotubes to make extremely dense electronic circuitry and the last year has seen major advances in creating basic electronic structures from nanotubes in the lab, from transistors up to simple logic elements. The gulf between these experiments and commercial nanotube electronics is, however, vast. There are various ways of producing SWNTs, which are briefly discussed later. The detailed mechanisms responsible for nanotube growth are still not fully understood and computer modeling is playing an increasing role in fathoming the complexities. The ambition of SWNT producers is to gain greater control over their diameters, lengths, and other properties, such as chirality (explained below). The volumes of SWNTs produced are currently small and the quality and purity are variable. Carbon Nanotechnologies Inc. of Houston, Texas, are currently ramping up production to a half a kilogram a day, which is actually huge in comparison to amounts of SWNTs that have been made historically. Various companies pursuing specific nanotube applications produce their own material in house. Imagine again the chicken wire that we roll up to make the nanotube. In fact, imagine a chicken wire fence out of which will be cut a rectangle to roll into a tube. You could cut the rectangle with the sides vertical or at various angles. Additionally, when joining the sides together, you can raise or lower one side. In some cases it will not be possible to make a tube such that the loose ends match and hexagons are formed, but in other cases it will, and these represent the possible permutations of SWNTs. The possibilities are two forms in which a pattern circles around the diameter of the tube, often called zigzag and armchair (not the most intuitive of names, unfortunately, but they are now widely used), and a variety of forms in which the hexagons spiral up or down the tube with varying steepness, these being the chiral forms. There is theoretically an infinite variety of the latter, if you allow for infinite diameters of nanotubes.

Which of these forms a nanotube takes is the major determinant of its electrical properties, i.e. whether the tube is semiconducting or conducting. For a long time, the fact that all known production methods created a mix of types has been considered one of the hurdles to be overcome if the electronic properties are to be exploited. Claims have now been made that it is possible to produce only the semiconducting kind (specifically, the zigzag form). Additionally, there are approaches that can yield only semiconducting nanotubes from a mix of semiconducting and conducting ones. One such approach relies on vaporizing the conducting nanotubes with a strong electric current, leaving only the semiconducting kind behind. A more recent approach is simply to leave the mix of nanotubes lying around for a while—the metallic ones are oxidized and become semiconducting (the process can, of course, be speeded up). In the early part of 2001 there were also reports that nanotubes had been induced to form crystals, each of which contained just one type of nanotube. This would have been a nice separation method but the silence on this approach since then suggests that these results have not been duplicated. Multi-walled carbon nanotubes are basically like Russian dolls made out of SWNTs—concentric cylindrical graphitic tubes. In these more complex structures, the different SWNTs that form the MWNT may have quite different structures (length and chirality). MWNTs are typically 100 times longer than they are wide and have outer diameters mostly in the tens of nanometers. Although it is easier to produce significant quantities of MWNTs than SWNTs, their structures are less well understood than single-wall nanotubes because of their greater complexity and variety. Multitudes of exotic shapes and arrangements, often with imaginative names such as bamboo-trunks, sea urchins, necklaces or coils, have also been observed under different processing conditions. The variety of forms may be interesting but also has a negative side—MWNTs always (so far) have more defects than SWNTs and these diminish their desirable properties. Many of the nanotube applications now being considered or put into practice involve multi-walled nanotubes, because they are easier to produce in large quantities at a reasonable price and have been available in decent amounts for much longer than SWNTs. In fact one of the major manufacturers of MWNTs at the moment, Hyperion Catalysis, does not even sell the nanotubes directly but only pre-mixed with polymers for composites applications. The tubes involved typically have 8 to 15 walls and are around 10 nanometres wide and 10 micrometers long. Other companies are moving into this space, notably formidable players like Mitsui, with plans to produce similar types of MWNT in hundreds of tons a year, a quantity that is greater, but not hugely so, than the current production of Hyperion Catalysis. This is an indication that even these less impressive and exotic nanotubes hold promise of representing a sizable market in the near future.

These are single-walled carbon cones with structures similar to those of nanotube caps that have been produced by high temperature treatment of fullerene soot. Sumio Iijima's group at NEC has demonstrated that nanohorns have good adsorptive and catalytic properties (i.e. desired substances stick to them and they enhance chemical reactions), and the company is working on using them in a new generation of fuel cells for personal electronics.

We use this term to refer to hollow and solid carbon fibers with lengths on the order of a few microns and widths varying from some tens of nanometers to around 200 nanometers. These materials have occasionally been referred to as nanotubes. However, they do not have the cylindrical chicken wire structure of SWNTs and MWNTs but instead consist of a mixture of forms of carbon, from layers of graphite stacked at various angles to amorphous carbon (lacking any large-scale regular structure). Because of this variable structure they do not exhibit the strength of pure nanotubes but can still be quite strong and possess other useful properties. The US company Applied Sciences Inc. is already producing tons of such material a year and there are several major producers in the Far East. Production processes for carbon nanotubes, crudely described, vary from blasting carbon with an electrical arc or a laser to growing them from a vapor, either en masse (usually in tangled bundles) or on nanoparticles, sometimes in predetermined positions. These processes vary considerably with respect to the type of nanotube produced, quality, purity and scalability. Carbon nanotubes are usually created with the aid of a metal catalyst and this ends up as a contaminant with respect to many potential applications, especially in electronics. IBM have very recently, however, grown nanotubes on silicon structures without a metal catalyst. Scalability of production processes is an essential commercial consideration—some of the approaches use equipment that simply cannot be made bigger and the only way to increase production is to make more pieces of equipment, which will not produce the economies of scale required to bring down costs significantly.

By now we hope you have an idea of the different types of nanotubes and how their qualities and ease of production vary. Reports of one company producing tons of the material, alongside reports of researchers not being able to get enough to get meaningful results from their research, should no longer be confusing. It would, of course, be useful if commentators got into the habit of making clear the type of material they are talking about. Understanding these differences is essential for understanding the commercial potential of the various applications of nanotubes and related structures that already exist or have been proposed. The variety of these is vast, and the commercialization timelines involved vary from now to ten years from now or more. Some of the potential markets are enormous. We will leave you with a taste of the possibilities. The materials markets are already seeing applications for composites based on multi-walled carbon nanotubes and nanofibers. In many ways this is an old market—that of carbon fibers, which have been around commercially for a couple of decades. The benefits of the new materials in these markets are the same as those of carbon fibers, just better; the main properties to be considered being strength and conductivity. Carbon fibers are quite large, typically about a tenth of a millimeter in diameter, and blacken the material to which they are added. MWNTs can offer the same improvements in strength to a polymer composite without the blackening and often with a smaller amount of added material (called the filler). The greater aspect ratio (i.e. length compared to diameter) of the newer materials can make plastics conducting with a smaller filler load, one significant application being electrostatic painting of composites in products such as car parts. Additionally, the surface of the composite is smoother, which benefits more refined structures such as platens for computer disk drives.

When thinking about structural applications such as these, it should be remembered that, in general, as the fibers get smaller so the number of defects decreases, in a progression towards the perfection of the single-walled nanotube. The inverse progression is seen in terms of ease of manufacturing—the more perfect, and thus more structurally valuable, the material, the harder it is to produce in quantity at a good price. This relationship is not written in stone—there is no reason that near-perfect SWNTs should not be producible cheaply and in large quantities. When this happens, and it might not be too far off, the improvements seen in the strength-to-weight ratios of composite materials could soar, impacting a wide variety of industries from sports equipment to furniture, from the construction industry to kitchenware, and from automobiles to airplanes and spacecraft (the aerospace industry is probably the one that stands to reap the greatest rewards). In fact a carbon nanotube composite has recently been reported that is six times stronger than conventional carbon fiber composites. This is an appropriate point at which to introduce a note of caution, and an area of research worth keeping an eye on. Just because the perfect nanotube is 100 times stronger than steel at a sixth of the weight doesn't mean you are going to be able to achieve those properties in a bulk material containing them. You may remember that the chicken wire arrangement that makes up the layers in graphite does not stick at all well to other materials, which is why graphite is used in lubricants and pencils. The same holds true of nanotubes—they are quite insular in nature, preferring not to interact with other materials. To capitalize on their strength in a composite they need to latch on to the surrounding polymer, which they are not inclined to do (blending a filler in a polymer is difficult even without these issues—it took a decade to perfect this for the new nanoclay polymers now hitting the markets).

One way of making a nanotube interact with something else, such as a surrounding polymer, is to modify it chemically. This is called functionalization and is being explored not just for composite applications but also for a variety of others, such as biosensors. For structural applications, the problem is that functionalization can reduce the valuable qualities of the nanotubes that you are trying to capitalize on. This is an issue that should not be underestimated. Of course, in theory you don't need to mix the nanotubes with another material. If you want to make super-strong cables, for example, the best solution would be to use bundles of sufficiently long nanotubes with no other material added. For this reason, one of the dreams of nanotube production is to be able to spin them, like thread, to indefinite lengths. Such a technology would have applications from textiles (the US military is in fact investigating the use of nanotubes for bullet-proof vests) to the 'space elevator'. The space elevator concept, which sounds like something straight out of science fiction (it was, in fact, popularized by Arthur C. Clarke) involves anchoring one end of a huge cable to the earth and another to an object in space. The taut cable so produced could then support an elevator that would take passengers and cargo into orbit for a fraction of the cost of the rockets used today. Sounds too far out? It has, in fact, been established by NASA as feasible in principle, given a material as strong as SWNTs. The engineering challenges, though, are awesome, so don't expect that 'top floor' button to be taking you into orbit any time soon.

The materials market is a big one, and there are others, which we'll come to, but smaller ones exist too. Nanotubes are already being shipped on the tip of atomic force microscope probes to enhance atomic-resolution imaging. Nanotube-based chemical and biosensors should be on the market soon (they face stiff competition from other areas of nanotechnology). The thermal conductivity of nanotubes shows promise in applications from cooling integrated circuits to aerospace materials. Electron beam lithography, which is a method of producing nanoscale patterns in materials, may become considerably cheaper thanks to the field emission properties of carbon nanotubes. Recent developments show promise of the first significant change in X-ray technology in a century. Entering more speculative territory, nanotubes may one day be used as nanoscale needles that can inject substances into, or sample substances from, individual cells, or they could be used as appendages for miniature machines (the tubes in multi-walled nanotubes slide over each other like graphite, but have preferred locations that they tend to spring back to). Big markets, apart from materials, in which nanotubes may make an impact, include flat panel displays (near-term commercialization is promised here), lighting, fuel cells and electronics. This last is one of the most talked-about areas but one of the farthest from commercialization, with one exception, this being the promise of huge computer memories (more than a thousand times greater in capacity than what you probably have in your machine now) that could, in theory, put a lot of the $40 billion magnetic disk industry out of business. Companies like to make grand claims, however, and in this area there is not just the technological hurdle to face but the even more daunting economic one, a challenge made harder by a host of competing technologies.

Despite an inevitable element of hype, the versatility of nanotubes does suggest that they might one day rank as one of the most important materials ever discovered. In years to come they could find their way into myriad materials and devices around us and quite probably make some of the leaders in this game quite rich.

The special nature of carbon combines with the molecular perfection of buckytubes (single wall carbon nanotubes) to endow them with exceptionally high material properties such as electrical and thermal conductivity, strength, stiffness, and toughness. No other element in the periodic table bonds to itself in an extended network with the strength of the carbon-carbon bond. The delocalised pi-electron donated by each atom is free to move about the entire structure, rather than stay home with its donor atom, giving rise to the first molecule with metallic-type electrical conductivity. The high-frequency carbon-carbon bond vibrations provide an intrinsic thermal conductivity higher than even diamond. In most materials, however, the actual observed material properties—strength, electrical conductivity, etc.—are degraded very substantially by the occurrence of defects in their structure. For example, high strength steel typically fails at about 1 per cent of its theoretical breaking strength. Buckytubes, however, achieve values very close to their theoretical limits because of their perfection of structure—their molecular perfection. This aspect is part of the unique story of buckytubes. Buckytubes are an example of true nanotechnology: only a nanometer in diameter, but molecules that can be manipulated chemically and physically. They open incredible applications in materials, electronics, chemical processing and energy management. Buckytubes have extraordinary electrical conductivity, heat conductivity and mechanical properties. They are probably the best electron field-emitter possible. They are polymers of pure carbon and can be reacted and manipulated using the tremendously rich chemistry of carbon. This provides opportunity to modify the structure and to optimise solubility and dispersion. Very significantly, buckytubes are molecularly perfect, which means that they are free of property-degrading flaws in the nanotube structure. Their material properties can therefore approach closely the very high levels intrinsic to them. These extraordinary characteristics give buckytubes potential in numerous applications.

Buckytubes are the best known field emitters of any material. This is understandable, given their high electrical conductivity, and the unbeatable sharpness of their tip (the sharper the tip, the more concentrated will be an electric field, leading to field emission; this is the same reason lightening rods are sharp). The sharpness of the tip also means that they emit at especially low voltage, an important fact for building electrical devices that utilize this feature. Buckytubes can carry an astonishingly high current density, possibly as high as 1013 A/cm^2. Furthermore, the current is extremely stable. An immediate application of this behaviour receiving considerable interest is in field-emission flat-panel displays. Instead of a single electron gun, as in a traditional cathode ray tube display, here there is a separate electron gun (or many) for each pixel in the display. The high current density, low turn-on and operating voltage, and steady, long-lived behaviour make buckytubes attract field emitters to enable this application. Other applications utilising the field-emission characteristics of buckytubes include: general cold-cathode lighting sources, lightning arrestors, and electron microscope sources.

Much of the history of plastics over the last half century has been as a replacement for metal. For structural applications, plastics have made tremendous headway, but not where electrical conductivity is required, plastics being famously good electrical insulators. This deficiency is overcome by loading plastics up with conductive fillers, such as carbon black and graphite fibres (the larger ones used to make golf clubs and tennis racquets). The loading required to provide the necessary conductivity is typically high, however, resulting in heavy parts, and more importantly, plastic parts whose structural properties are highly degraded. It is well established that the higher aspect ratio of filler, the lower loading required to achieve a given level of conductivity. Buckytubes are ideal in this sense, since they have the highest aspect ratio of any carbon fibre. In addition, their natural tendency to form ropes provides inherently very long conductive pathways even at ultra-low loadings. Applications that exploit this behaviour of buckytubes include EMI/RFI shielding composites and coatings for enclosures, gaskets, and other uses; electrostatic dissipation (ESD), and antistatic materials and (even transparent!) coatings; and radar-absorbing materials.

Buckytubes have the intrinsic characteristics desired in material used as electrodes in batteries and capacitors, two technologies of rapidly increasing importance. Buckytubes have a tremendously high surface area (~1000 m^2/g), good electrical conductivity, and very importantly, their linear geometry makes their surface highly accessible to the electrolyte. Research has shown that buckytubes have the highest reversible capacity of any carbon material for use in lithium-ion batteries. In addition, buckytubes are outstanding materials for supercapacitor electrodes and are now being marketed. Buckytubes also have applications in a variety of fuel cell components. They have a number of properties including high surface area and thermal conductivity that make them useful as electrode catalyst supports in PEM fuel cells. They may also be used in gas diffusion layers as well as current collectors because of their high electrical conductivity. Buckytubes' high strength and toughness to weight characteristics may also prove valuable as part of composite components in fuel cells that are deployed in transport applications where durability is extremely important.

The same issues that make buckytubes attractive as conductive fillers for use in shielding, ESD materials, etc., make them attractive for electronics materials, such as adhesives and other connectors (e.g., solders). The idea of building electronic circuits out of the essential building blocks of materials—molecules—has seen a revival the past five years, and is a key component of nanotechnology. In any electronic circuit, but particularly as dimensions shrink to the nanoscale, the interconnections between switches and other active devices become increasingly important. Their geometry, electrical conductivity, and ability to be precisely derived, make buckytubes the ideal candidates for the connections in molecular electronics. In addition, they have been demonstrated as switches themselves.

The record-setting anisotropic thermal conductivity of buckytubes is enabling applications where heat needs to move from one place to another. Such an application is electronics, particularly advanced computing, where uncooled chips now routinely reach over 100°C. CNI's technology for creating aligned structures and ribbons of buckytubes is a step toward realising incredibly efficient heat conduits. In addition, composites with buckytubes have been shown to dramatically increase the bulk thermal conductivity at small loadings.

The world-record properties of buckytubes are not limited to electrical and thermal conductivities, but also include mechanical properties, such as stiffness, toughness, and strength. These properties lead to a wealth of applications exploiting them, including advanced composites requiring high values in one or more of these properties.

Fibres spun of pure buckytubes have been demonstrated and are undergoing rapid development, along with buckytube composite fibres. Such super strong fibres will have applications including body and vehicle armour, transmission line cables, woven fabrics and textiles.

Buckytubes have an intrinsically high surface area; in fact, every atom is not just on a surface—each atom is on two surfaces, the inside and outside! Combined with the ability to attach essentially any chemical species to their sidewalls provides an opportunity for unique catalyst supports. Their electrical conductivity may also be exploited in the search for new catalysts and catalytic behaviour. The exploration of buckytubes in biomedical applications is just underway, but has significant potential. Cells have been shown to grow on buckytubes, so they appear to have no toxic effect. The cells also do not adhere to the buckytubes, potentially giving rise to applications such as coatings for prosthetics and anti-fouling coatings for ships. The ability to chemically modify the sidewalls of buckytubes also leads to biomedical applications such as vascular stents, and neuron growth and regeneration. There is a wealth of other potential applications for buckytubes, such as solar collection; nanoporous filters; catalyst supports; and coatings of all sorts. There are almost certainly many unanticipated applications for this remarkable material that will come to light in the years ahead and which may prove to be the most important and valuable of all.

Scanning Probe Technology

The third general pathway leading to molecular manufacturing involves a technology known as scanning probe microscopes (SPMs). The first of the SPMs was the Scanning Tunneling Microscope (STM) developed in the late 1970s and early 1980s by Gerd Karl Binnig and Heinrich Rohrer at an IBM research lab in Zurich, Switzerland, earning these scientists, along with Ernst Ruska, the 1986 Nobel in Physics. The STM was initially used as an imaging device, capable of resolving individual atoms by recording the quantum tunneling current that occurs when an extremely sharp conductive probe tip (usually tungsten, nickel, gold, or PtIr) is brought to within about one atomic diameter of an atom, and then adjusting the position of the tip to maintain a constant current as the tip is scanned over a bumpy atomic surface. A height change as small as 0.1 nm can cause tunneling current to double. The tip is connected to an arm that is moved in three dimensions by stiff ceramic piezoelectric transducers that provide subnanometer positional control. If the tip is atomically sharp, then the tunneling current is effectively confined to a region within ~0.1 nm of the point on the surface directly beneath the tip, thus the record of tip adjustments generates an atomic-scale topographic map of the surface. STM tips can scan samples at ~KHz frequencies, although slower scans are used for very rough surfaces. In some modern STMs (e.g., the DI Nanoscope), the sample is moved while the tip is held stationary.

A major limitation of the STM was that it only worked with conducting materials such as metals or semiconductors, but not with insulators or biological structures such as DNA. To remedy this situation, in 1986 Binnig, Quate and Gerber developed the

Atomic Force Microscope (AFM) which is sensitive directly to the forces between the tip and the sample, rather than a tunneling current. An AFM can operate in at least three modes. In "attractive" or non-contact mode (NC-AFM, 0.01-1 N/m force constant), the tip is held some tens of nanometers above the sample surface where it experiences the attractive combination of van der Waals, electrostatic, or magnetostatic forces. In "repulsive" or contact mode (CAFM, 0.01-1 N/m force constant), the tip is pressed close enough to the surface for tip and sample electron clouds to overlap, generating a repulsive electrostatic force of ~10 nN, much like the stylus riding a groove in a record player. There is also intermittent-contact mode (IC-AFM, 0.01-1 N/m force constant), which is sometimes called "tapping" mode. In any of these modes, a topographic map of the surface is generated by recording the up-and-down motions of the cantilever arm as the tip is scanned. These motions may be measured either by the deflection of a light spot reflected from a mirrored surface on the cantilever or by tiny changes in voltage generated by piezoelectric transducers attached to the moving cantilever arm. Typical AFM cantilevers have lengths of 100-400 microns, widths of 20-50 microns, and thicknesses between 0.4 to several microns. AFM tips may be positioned with ~0.01 nm precision, compressive loads as small as 1-10 pN are routinely measured, and the tips may be operated even in liquids.

S. Vetter notes that STM technology has also improved, reaching resolutions of ~0.001 nm in the z direction (vertical) and ~0.01 nm in the xy plane, well beyond atomic resolution. The STM remains the instrument with the best resolution. The conducting surface limitation has been overcome in some cases by coating the target with an extremely thin conducting layer, developing tips with multiple electrodes, and coating a conducting substrate with a sample so thin as to allow enough conduction even if the sample is characterized as an insulator in bulk. By 1998, the growing family of SPMs included at least forty types of instruments and techniques that relied on interactions between a scanned surface and a nearby probe. Different instruments measured different forces and thus could be used to characterize different properties of the surface. For example, friction force microscopes (FFMs), magnetic force microscopes (MFMs), shear force microscopes (ShFMs), scanning capacitance microscopes (SCMs), scanning conducting ion microscopes, chemical force microscopes, and electrostatic force microscopes (EFMs) measured frictional drag or other binding forces. Magnetic resonance force microscopes (MRFM) used a field generated from a small magnet mounted on the tip of the cantilever arm to probe nuclear magnetic moments in a small region on the surface of the sample, imaging atom types and even detecting the spin of a single electron. By the mid-1990s, AFMs were already a $100 million/year industry and SPMs generally were an off-the-shelf technology costing up to $50,000-$500,000 for complete systems, with the whole industry worth up to ~$0.5 billion annually. (Low-performance "homebrew," student, and science-fair STMs have been built for as little as $50.)

How might SPMs be used for molecular manufacturing? In the crudest approach, SPMs might be employed as nanoscale milling machines to carve out "nanoparts" from appropriate substrates. For example, Kim and Lieber used an AFM to perform nanomachining operations on a molybdenum trioxide crystal — applying a 100 nN load at the tip, they milled a triangular-shaped 50-nm planar chunk from the crystal, then slid the part 200 nm across the work surface by pushing it with the AFM tip. Sheehan and Lieber milled a 50-nm rectangular part and two other parts with concavities complementary to the rectangle by mechanically etching a MoO_3 layer deposited atop an MoS_2 surface by pressing down with an AFM tip at a 50 nN load. Like a numerical-controlled machine tool, the AFM motions could be programmed to perform a series of steps, like automatically carving the chemical formula "MoO_3" into the crystal. Nanoparts with features as narrow as 10 nm can be nanomilled in a limited set of materials [P.E. Sheehan, personal communication, 1995]. In theory, large arrays of independently-controlled SPM tips could be operated in parallel to carve out great numbers of similar objects simultaneously, though such objects would of course not be atomically identical without further "finishing." AFM tips have also been used to bulldoze nanoscale lines and rectangular <10-nm features into nonconducting photoresist, perform "dip-pen" AFM nanolithography, form grooves, and fabricate a single-electron transistor via an STM nano-oxidation process. A carbon nanotube affixed to an AFM tip was used as a pencil to write 10-nm-wide features onto silicon substrates at ~0.5 mm/sec: An electric field issues from the nanotube, removing hydrogen atoms from a hydrogen monolayer atop a silicon base; the exposed silicon surface oxidized, producing narrow SiO_2 tracks.

Perhaps more significantly, SPMs have been employed in an increasingly sophisticated manner to manipulate individual atoms and small clusters of atoms. The possibility of modifying surfaces with scanning probe tips was evident from the earliest years of STM research, since inadvertent contact between tips and surfaces routinely caused such modifications, and the possibility of transferring material between tip and sample had already been discussed. Suggestions for controlled surface modification soon appeared, and in 1985 Becker and Golovchencko used voltage pulses on an STM tip to pluck a single germanium atom from the {111} surface of a sample. In 1988, J. Foster and colleagues at IBM Almaden pinned small organic molecules to a graphite surface by applying a small electrical pulse through an STM. Additional pulses enlarged or erased this molecular feature, and often split the organic molecule into smaller pieces, though not in a controlled manner. Other workers used an STM to dump small clusters of gold atoms on a platinum surface, clusters of copper atoms on a gold surface, and molecules of tungsten carbide from a supply flowing over a surface. Others reported positioning carbon monoxide molecules and platinum atoms on platinum surfaces, moving clusters and single atoms of silicon across a silicon surface at room temperature, and fabricating via STM a "molecular corral" or "quantum corral" — a ring of atoms so small that the enclosed electrons were forced to exhibit quantum behavior. In 1999, Tomanek and Kral proposed an atomic fountain pen in which a carbon nanotube filled with atoms is electrically induced to release these atoms, one by one every 15 microsec, onto the work surface.

More precise control was achieved by Eigler and Schweizer at IBM Almaden in 1989, when they used an STM to position 35 individual xenon atoms on a nickel surface to spell out the corporate logo "IBM". To accomplish this, a bias voltage was applied to weaken the adsorption of each atom to its nickel substrate, then each atom was dragged, one by one, to the desired locations at a speed of ~0.4 nm/sec to build a meaningful pattern. It took 22 hours to make the entire logo, or ~38 min/atom. The experiment had to be carried out at 4 K (i.e., liquid helium temperatures) because the arrangement would have been unstable at room temperature. Numerous similar examples of "atomic graffiti" followed the IBM experiment, including atoms patterned in the shape of a world map; the word "atom" and "nanoworld", spelled out with atoms, in Japanese; the word "Peace," Einstein's famous equation "E = mc," and even a reproduction of the well-known Einstein portrait with the scientist's tongue sticking out; a sketch of a "molecule man" using 28 individual CO molecules on a platinum surface; gold nanoparticles spelling out "USC" and "Zyvex"; and sulfur atoms removed from a surface to make letters 2 nm high. In 1996, Gimzewski's group used an STM tip to manipulate individual buckyballs along terraces on a grooved copper plate, making a "molecular abacus."

Some variant of an SPM would appear to be the most promising tool for direct molecular manipulation. Once a tip lies within ~1 nm of a surface, the potential barrier can be lowered sufficiently for atoms to be induced from the surface by field evaporation; an applied tip voltage ionizes the atom which can then be guided around by the tip. Depending on voltage and separation, atoms, atom clusters, or molecules can be pried out, pushed in, or nudged around on a surface. By 1998, techniques for direct atomic manipulation proceeded at room temperature and ~1000 times faster than the original IBM logo had been assembled. For example, the Nanomanipulator, an interactive haptic control system created by a group at the University of North Carolina at Chapel Hill, allows near-real-time manipulation of individual gold atoms across a surface using a hand-held master-slave controller that drives an STM probe while the position of the atom is displayed on a monitoring screen visible to the user. Partially funded by NSF, Silicon Graphics, and TopoMetrix, the Nanomanipulator is a fully-integrated system that enables investigators to "feel" the interatomic forces as the user pushes atoms around on a surface. In one demonstration, a TV-watching user herded a gold atom, slipping and sliding, into a slot in the planar workplace in about 1 minute; when force feedback was turned off, movements tended to run wild. IBM Almaden has also experimented with removing individual atoms from metal surfaces using an STM hooked into a Virtual Reality Dataglove apparatus.

To manufacture atomically precise parts, it also will be necessary to manipulate covalent bonds at the probe tip. STMs have broken and created chemical bonds; ~1 volt pulses were used to pull atoms out of crystals, binding them to the tip, and then to reinsert them back into the crystal. In 1995, the first demonstration of catalysis on a nanometer scale was reported by scientists at the Molecular Design Institute at LBL.

They used an AFM modified to function like an ultrafine-point pen for catalytic calligraphy to change the chemical composition of a material surface one molecule at a time. A surface was prepared as a self-assembled monolayer (SAM) of alkylazide molecules capped with a crown of three nitrogen atoms, then platinum-coated chromium was deposited onto an AFM silicon tip just a few atoms wide. The SAM was soaked with a hydrogen-containing solvent, then scanned by the AFM over a 100 micron^2 area, with the platinum catalyzing a covalent bonding reaction in which hydrogen was added to the azides, transforming them into amines as revealed by selective fluorescent tags.

Tip technology is an important and fast-growing subspecialty of SPM research. In 1990, Drexler and Foster suggested the use of custom-made, synthetic proteins to be mounted on the point of an SPM tip. Modified antibodies or specially designed proteins could serve as simple, first-generation "grippers" for binding and manipulating specific molecules, bringing them into position to react with other molecules in a precise and selective way. Functionalized SPM tips have been created to exploit antibody-antigen recognition and in the context of chemical force microscopy. A major convenience for molecular manipulation is that the same instrument that does the chemistry and the molecular orienting can also be used to inspect the results.

In 1996, Smalley's group at Rice University attached a single nanotube to the pyramidal silicon tip of an AFM and showed that it was quite robust and could image the bottom of deep trenches inaccessible to conventional tips. By 1998, progress had been made in functionalized carbon nanotubes for use on AFM tips. Most notably, Wong and colleagues prepared nanotube tips by oxidation in air at 700°C, burning off all but 2% of the original material and leaving the ends covered with carboxyl (COOH) groups whose chemistry is rich and well understood. Four different kinds of tips were created: (1) the original carboxyl tip, which is acidic; (2) an amine-terminated tip (made by forming an amide bond to one of the amine groups in ethylenediamine ($H_2NCH_2CH_2NH_2$)), which is basic; (3) a hydrocarbon-terminated tip (made by forming an amide bond to benzylamine ($C_6H_5CH_2NH_2$)), which is hydrophobic; and (4) a biotin-terminated tip (made by forming an amide bond to a biotin derivative), which shows specific binding to streptavidin. AFM contact forces between tips and selected substrates were shown to be sensitive to pH and to the chemical details of the substrate in ways consistent with the tips' intended chemistry.

Wong's tips had three closely related advantages over previous techniques. First, when a functional group is attached to the apex of an Si_3N_4 or SiO_2 tip, these groups usually adhere to the sides of the tip as well; upon using the tip as a tool, there is a constant hazard that contact with the sides of the tip will alter the workpiece in unwanted places. Unlike these tips built with bulk techniques, the ends of Wong's tips are very different from their sides — the carboxyl groups are attached only on the ends of the nanotube, not on the sides. Second, Wong's tips have lateral dimensions set by the nanoscale dimensions of nanotubes, not by top-down fabrication techniques. Note the authors: "...the small effective radius of nanotube tips significantly improves resolution beyond what can be achieved using commercial silicon tips....we have recently demonstrated that lateral resolution of <3 nm can be achieved by using COOH-terminated single-walled nanotubes tips". Third, single-walled nanotube tips are much closer to yielding truly single atom tips with controlled chemistry than any other alternative. J. Soreff notes that a (10, 10) nanotube with a 1.4-nm diameter has just 20 atoms at an open end. Even given a statistical distribution of tubes with varying numbers of carboxyl groups attached to their ends, one should be able to build a ligand which covers the whole end of the tube, yielding a method for ensuring that just one molecule of known structure and orientation was present at the end of the probe.

Theoretical studies have employed molecular dynamics simulations and other techniques to investigate the behavior of mechanosynthetic tool tips. For example, Drexler proposed using an acetylene radical to perform selective abstractions of hydrogen from a diamond surface, and this proposal was further studied in Musgrave's ab initio quantum chemistry analysis and Sinnott's molecular dynamics modeling of the reaction at room temperature on a diamond {surface; other H-abstraction studies have been done. One possible structure for a hydrogen abstraction tool might be similar to an anthracene whose end had been modified as shown; prior to use, the structure must be activated by removal of the terminal hydrogen. Brenner and colleagues describe further work modeling the parallel abstraction of several hydrogen atoms from the diamond surface. Such reactions are similar to those involved in the well-studied chemical vapor deposition of diamond, but adding only positional control and applied mechanical force; more detailed theoretical studies of diamond mechanosynthesis reactions have begun.

R. Merkle has proposed a simple mechanosynthetic system capable of fabricating a large class of useful stiff hydrocarbons. Merkle's system includes positionally-controlled hydrogen abstraction and deposition tools (to remove or add hydrogen atoms to a workpiece), carbene and dimer deposition tools (to add one or two carbons to a workpiece), a silicon radical tool (to remove a carbon), and four other specialized tools. Each proposed tool has a molecular handle structure that could be permanently affixed to a suitably functionalized AFM tip (or other appropriate mechanosynthetic nanomanipulator). As each operation is performed on the butadiyne (a.k.a diacetylene, C_4H_2) hydrocarbon feedstock in an inert (e.g., vacuum or noble gas) environment, one or more tools are brought to the workpiece at the necessary position and angle, and any necessary force is applied via the AFM. Interestingly, except for a small number of "vitamin parts" involving transition metals, Merkle's system can in theory synthesize all of its own mechanosynthetic tools, thus establishing net positive production of system components and illustrating a limited subset-level "quantitative parts closure" as first defined in 1982 by Freitas and Gilbreath in connection with *self-replicating machine systems*. Merkle acknowledges that several of the proposed reactions involve the simultaneous coordinated action of two, three, or even four positionally controlled tools, but suggests that two 6-degree-of-freedom manipulators plus some combination of appropriate jigs and fixtures should be sufficient in a more parsimonious system. However, steric hindrance of multiple tools at one site also was not explicitly addressed. Another potential problem to be overcome is loss of positional registration as the various functionalized tool heads are changed out at the AFM tip; such tips might be paired with a second positionally-invariant metrology tip on each tool head, allowing precise calibration of workspace location before proceeding back to the workpiece.

Others have suggested fabricating nanoparts by building up structures by stacking layers of individual atoms, perhaps constrained by sacrificial joiners and scaffolding, and honed by differential etching operations. Assuming that "nanoparts" can be fabricated to atomic precision, can SPMs also be used to assemble these nanoparts into working nanomachines? Some preliminary work has been accomplished. For instance, arrays of nanoscale holes have been synthesized that could serve as "workpiece holders." In one experiment, C_{60} buckyballs were scattered about on a prepared surface, then an AFM pushed the individual spheres into their own little holes, securely seating them pending further processing. The C_{60}-impregnated perforated film was so durable that the buckyballs remained trapped after several months of storage. Similarly, Jung and colleagues in Gimzewski's group synthesized 4-legged porphyrin-based molecular "nanoparts" that were specifically designed for easy positioning on a copper surface. Pushing with an STM at room temperature, the authors displaced the ~1.5-nm wide nanoparts in predefined directions and rotated the 4-legged molecules at will, arranging, for example, the parts into a precise hexagonal configuration on the copper surface. The STM has been used to induce, and to view, the rotation of an individual oxygen molecule trapped on a platinum surface, as it is intentionally and stably rotated into any one of three distinct orientations. The forces required to slide, or alternatively to roll, a carbon nanotube across a graphitic surface have been directly measured by AFM.

The first known instance of a crude but purely mechanical "nanopart" assembly operation was performed in 1995 by Paul E. Sheehan and Charles M. Lieber. Two nanoscale parts with rectangular slots and a third 50-nm rectangular sliding "latch" member were milled from a MoO_3 crystal using an AFM, then the rectangular "latch" member was slid repeatedly from one slot to the other using the same AFM, making a three-nanopart reversible mechanical latch. Noted the authors: "The lateral force needed to break the latch, 41 nN, was large considering the small latch contact area, which suggests that relatively robust assemblies can be created with such devices. Such a reversible latch could serve as the basis for mechanical logic gates....Our results...demonstrate the ability to machine complex shapes and to reversibly assemble these pieces into interlocking structures."

In 1998, manipulation of nanoscale parts in the vertical dimension (e.g., normal to the surface plane) had only just begun. The first three-dimensional structure built out of single "nanoparts" was demonstrated in 1997 by Requicha's group at the USC Laboratory for Molecular Robotics. First, an AFM was used to push 5-nm gold nanoparticles in the third dimension up and over surface protrusions or obstructions — for example, a 5-nm gold particle was shoved up onto a 2-nm-high protrusion, then slid back off again onto the surface. In another series of experiments conducted in air at room temperature with 15-30 nm gold nanoparticles deposited on silicon previously coated with a silane layer, Requicha's group used an AFM to build a simple 3-D pyramidal structure by pushing one gold "nanopart" on top of two others, then off again; the group also used the AFM to rotate and translate a dimer unit formed by two linked "nanoparts".

Three-dimensional nanoassembly might perhaps be more readily achieved if nanoparts could assist in their own assembly process. R. Merkle [personal communication, 1998] suggests that an SPM could be used to position DNA-tagged molecular building blocks. A building block tagged with a specific single-stranded DNA should preferentially adhere to a surface covered with the complementary DNA. If the building block has a second single-stranded DNA tag which is complementary to single-stranded DNA on the tip of an SPM, and if it is initially attached to the SPM tip by this tag, then it could be positioned by the SPM and attached to the surface. The relative strength of attachments could be adjusted by changing the number of base pairs in the complementary region between the two strands. A related suggestion, due to G. Fahy, is that SPM tips for assembly could be designed with more than one binding site. The existence of at least two binding sites could allow precise orientation of a nanopart, permitting the nanopart to be presented to a desired target site (or another nanopart, possibly attached to another manipulator) at exactly the location and in exactly the orientation desired, assuming that each nanopart can bind to each tip in only one possible orientation. A convenient post-assembly nanopart release mechanism is also required.

In 1997, the Avouris group at IBM Yorktown Heights demonstrated that the tip of an atomic force microscope (AFM) could be used to control the shape and position of individual multiwalled carbon nanotubes dispersed on a surface. Nanotubes could be bent, straightened, translated, rotated, and (under certain conditions) cut. In 1998, Zyvex LLC (a developmental engineering company whose goal is to create a molecular assembler capable of manufacturing atomically precise structures) demonstrated the ability to manipulate carbon nanotubes in three dimensions inside a scanning electron microscope (SEM) having ~6 nm resolution at near-video scan rates. Zyvex's custom piezoelectric vacuum manipulator achieved positional resolutions comparable to SPMs along with the ability to manipulate objects along one rotational and three linear degrees of freedom with 0.1 nm spatial resolution. This prototypical device could probe, select and handle limited classes of nanometer-scale objects such as carbon nanotubes under real-time SEM inspection. Carbon nanotubes were attached to commercial atomic force microscope (AFM) tips either by van der Waals forces alone or by "nanosoldering" by a concentrated electron beam from the SEM. Forces applied to the nanotubes could be measured from the cantilevers' deflections (spring constants 0.01-100 N/m; see also Neumister and Ducker). Sometimes nanotubes were transferred from tip to tip. The manipulator could function both as a research tool for investigating properties of carbon nanotubes and other nanoscale objects without surface restrictions, and as a rudimentary building device for larger nanotube assemblies. Zyvex believed that this capability to select and manipulate nanoscale components and to examine directly their suitability as construction materials during various phases of the construction process would play an important role in enabling the technology of assembling mechanical and electronic devices from prefabricated components. Most impressively, the Zyvex system allowed the simultaneous coordinated operation of three independently controlled AFM tips within the same workspace, at different orientations. (A four-tip system was expected to be operational by mid-1999 [Mark Dyer, personal communication, 1999].) Two-handed coordinated micromanipulation under the view of an SEM had previously been demonstrated by others.

Tuzun and colleagues calculated the requirements for coaxial docking of two nanotubes of different diameters to form a molecular bearing. They looked at bearings formed from nanotubes 11 rings long, with 10 carbons per ring in the shaft and either 30 or 34 carbons per ring in the sleeve. For the computer simulations, Tuzun placed the sleeve of the bearing along the z axis and gave the shaft a small initial velocity towards it. A perfectly aligned shaft falls straight into a potential well from the van der Waals attraction to the sleeve; displaced or misoriented shafts can bounce off the edge of the sleeve. The docking envelopes for the molecular dynamics calculations and for the rigid body calculations had essentially the same shape, but with slightly different sizes. For the atomistic calculation, a shaft aligned parallel to the sleeve (with a 30 carbon ring sleeve) can be displaced by 0.26 nm before it fails to dock, while in the rigid body calculation the displacement can only be 0.16 nm. The authors note that "if an end of the shaft points closely enough to the center of the sleeve, it will fall into the nonbonded potential energy well and the two nanotubes will dock." These calculations are important because they tell us what tolerances can be accepted during the manual assembly process. Note the authors: "A question just as important as how to design or operate nanomachines is how to assemble them". Interestingly, one of Zyvex's experiments produced a multiwalled carbon nanotube that thinned in three steps along its length, suggesting the most likely explanation that the inner shells had been pulled from the outer shells. This phenomenon has also been observed by others and is known as the "sword and sheath" failure — the exact inverse of the purposeful nanotube insertion operation investigated by Tuzun and colleagues.

In order to produce large numbers of nanoparts and nanoassemblies, massively parallel SPM arrays and microscale SPMs would be most convenient. Force sensing devices such as piezoelectric, piezoresistive, and capacitive micro-cantilevers made it possible to construct microscale AFMs on chips without an external deflection sensor. (We exclude fixed-tip arrays in the following discussion.) In 1995, Itoh and colleagues at the University of Tokyo fabricated an experimental piezoelectric ZnO_2onSiO_2 microcantilever array of ten tips on a single silicon chip. Each cantilever tip lay ~70 microns from its neighbor, and measured 150 microns long, 50 microns wide and 3.5 microns thick, or ~26,000 micron/device, and each of the devices could be operated independently in the z-axis (e.g., vertically) up to near their mechanical resonance frequencies of 145-147 KHz at an actuation sensitivity of ~20 nm/volt — for instance, 0.3-nm resolution at 125 KHz.

Parallel probe scanning and lithography has been achieved by Quate's group at Stanford, which has progressed from simple piezoresistive microcantilever arrays with 5 tips spaced 100 microns apart and 0.04-nm resolution at 1 KHz but only one z-axis actuator for the whole array, to arrays with integrated sensors and actuators that allow parallel imaging and lithography with feedback and independent control of each of up to 16 tips, with scanning speeds up to 3 mm/sec using a piezoresistive sensor. By 1998, Quate's group had demonstrated arrays of 50-100 independently-controllable AFM probe tips mounted in 2-D patterns with 60 KHz resonances, including a 10 x 10 cantilevered tip array fabricated in closely spaced rows using throughwafer interconnects on a single chip.

MacDonald's group at the Cornell Nanofabrication Facility has pursued similar goals. In 1991, the team fabricated their first submicron stylus, driven in the xy plane using interdigitating MEMS comb drives, including the first opposable tip pair. By 1993, they had produced a 25-tip array on one xyz actuator, and by 1995 a complete working microSTM (including xy comb drives) measuring 200 microns on an edge and a microAFM measuring 2 mm on an edge including a 1-mm long cantilever with a 20-nm diameter integrated tip on a 6-micron high by 1-micron diameter support shaft. MacDonald's group demonstrated tip arrays with 5 micron spacings, exploiting the same process used to make the working microSTM. With the same technology tips or small arrays of tips could be spaced 25-50 microns apart and integrated with individual z-axis microactuators, so that one xy-axis manipulator could support many tips with each tip having a separate z actuator. By 1997, the group had built and tested an array of microSTMs on the surface of an ordinary silicon chip, with each tip on a cantilever 150 microns long with 3-D sensing and control. The largest prototype array has 144 probes, arranged in a square consisting of 12 rows of 12 probes each, with individual probe needles about 200 microns apart. Further development was to focus on increasing the range of movement and on fitting more and smaller probes into the same space.

Gene Therapy for Plants Using Carbon Nanofibres

Researchers are developing new techniques that use nanoparticles for smuggling foreign DNA into cells. For example, at Oak Ridge National Laboratory, the US Department of Energy lab that played a major role in the production of enriched uranium for the Manhattan Project, researchers have hit upon a nano-technique for injecting DNA into millions of cells at once. Millions of carbon nanofibres are grown sticking out of a silicon chip with strands of synthetic DNA attached to the nanofibres. Living cells are then thrown against and pierced by the fibres, injecting the DNA into the cells in the process.

"It's like throwing a bunch of baseballs against a bed of nails...We literally throw the cells onto the fibers, and then smash the cells into the chip to further poke the fibers into the cell." - Timothy McKnight, engineer, Oak Ridge Laboratory

Once injected, the synthetic DNA expresses new proteins and new traits. Oak Ridge has entered into collaboration with the Institute of Paper Science and Technology in a project aimed to use this technique for genetic manipulation of loblolly pine, the primary source of pulpwood for the paper industry in the USA.

Unlike existing genetic engineering methods, the technique developed by Oak Ridge scientists does not pass modified traits on to further generations because, in theory, the DNA remains attached to the carbon nanofibre, unable to integrate into the plants' own genome. The implication is that it would be possible to reprogram cells for one time only. According to Oak Ridge scientists, this relieves concerns about gene flow associated with genetically modified plants, where genes are transferred between unrelated organisms or are removed or rearranged within a species.

If the new technique enables researchers to selectively switch on or off a key trait such as fertility, will seed corporations use the tiny terminators to prevent farmers from saving and re-using harvested seed - compelling them to return to the commercial seed market every year to obtain the activated genetic trait they need?

This approach also raises a number of safety questions: what if the nanofibres were ingested by wildlife or humans as food? What are the ecological impacts if the nanofibres enter the cells of other organisms and cause them to express new proteins? Where will the nanofibres go when the plant decomposes in the soil? Carbon nanofibres have been compared to asbestos fibres because they have similar shapes. Initial toxicity studies on some carbon nanofibres have demonstrated inflammation of cells. A study by NASA found inflammation in the lungs to be more severe than in cases of silicosis, though Nobel laureate Richard Smalley, Chairman of Carbon Nanotechnologies Inc. gives little weight to these concerns: "We are confident there will prove out to be no health hazards but this [toxicology] work continues."

Nanotech Applications

Laboratories throughout the world are rapidly gaining atomically precise control over matter. As this control extends to an ever wider variety of materials, processes and devices, opportunities for applications relevant to NASA's missions will be created. This document surveys a number of future molecular nanotechnology capabilities of aerospace interest. Computer applications, launch vehicle improvements, and active materials appear to be of particular interest. We also list a number of applications for each of NASA's enterprises. If advanced molecular nanotechnology can be developed, almost all of NASA's endeavors will be radically improved. In particular, a sufficiently advanced molecular nanotechnology can arguably bring large scale space colonization within our grasp. This study describes potential aerospace applications of molecular nanotechnology, defined as the thorough three-dimensional structural control of materials, processes and devices at the atomic scale. The inspiration for molecular nanotechnology comes from Richard P. Feynman's 1959 visionary talk at Caltech in which he said, "The problems of chemistry and biology can be greatly helped if our ability to see what we are doing, and to do things on an atomic level, is ultimately developed—a development which I think cannot be avoided." Indeed, scanning probe microscopes (SPMs) have already given us this ability in limited domains. Synthetic chemistry, biotechnology, "laser tweezers" and other developments are also bringing atomic precision to our endeavors. An expanded version of Drexler's MIT PhD thesis, examines one vision of molecular nanotechnology in considerable technical detail. It proposes the development of programmable molecular assembler/replicators. These are atomically precise machines that can make and break chemical bonds using mechanosynthesis to produce a wide variety of products under software control, including copies of themselves. Interestingly, living cells exhibit many properties of assembler/replicators. Cells make a wide variety of products, including copies of themselves, and can be programmed with DNA. Replication is one approach to building large systems, such as human rated launch vehicles, from molecular machines manipulating matter one or a few atoms at a time. Note that biological replication is responsible for systems as large as redwood trees and whales.

Another approach to nanotechnology is supramolecular self-assembly, where molecular systems are designed to attract each other in a particular orientation to form larger systems. Hollow spheres large enough to be visible in a standard light microscope have been created this way using self-assembling lipids. There are many other examples and this field is rapidly advancing. Biological systems can do most of what molecular nanotechnology strives to accomplish — atomically precise products, active materials, reproduction, etc. However, biological systems are extremely complex and molecular nanotechnology seeks simpler systems to understand, control and manufacture. Also, biological systems usually work at fairly mild temperature and pressure conditions in solution — conditions that are not found in most aerospace environments. Today, extremely precise atomic and molecular manipulation is common in many laboratories around the world and our abilities are rapidly approaching Feynman's dream. The implications for aerospace development are profound and ubiquitous. A number of applications are mentioned here and a few are described in some detail with references. From this sample of applications it should be clear that although molecular nanotechnology is a long term, high risk project, the payoff is potentially enormous — vastly superior computers, aerospace transportation, sensors and other technologies; technologies that may enable large scale space exploration and colonization.

This chapter is organized into two sections. In the first, we examine three technologies—computers, aerospace transportation, and active materials—useful to nearly all NASA missions. In the second, we investigate some potential molecular nanotechnology payoffs for each area identified in NASA's strategic plan. Some of these applications are under investigation by nanotechnology researchers at NASA Ames. Some of the applications described below have relatively near-term potential and working prototypes may be realized within three to five years. This is certainly not true in other cases. Indeed, many of the possible applications of nanotechnology that we describe here are, at the present time, rather speculative and futuristic. However, each of these ideas have been examined at least cursorily by competent scientists, and as far as we know all of them are within the bounds of known physical laws. We are not suggesting that their achievement will be easy, cheap or near-term. Some may take decades to realize; some other ideas may be scrapped in the coming years as insuperable barriers are identified. But we feel that they are worth mentioning here as illustrations of the potential future impact of nanotechnology.

The applicability of manufacturing at an ever smaller scale is nowhere more self-evident than in computer technology. Indeed, Moore's law (an observation not a physical law) says that computer chip feature size decreases exponentially with time, a trend that predicts atomically precise computers by about 2010-2015. This capability is being approached from many directions. Here we will concentrate on those under development by NASA Ames and her partners.

Carbon nanotubes can be viewed as rolled up sheets of graphite from 0.7 to many nanometers in diameter. The smaller tubes are single molecules. Dai placed carbon nanotubes on an SPM tip thus extending our ability to manipulate a single molecule with sub-angstrom accuracy. Not only are the tips atomically precise, but they should have approximately the same chemistry as C_{60}, and thus be functionalizable with a wide variety of molecular fragments. Functionalizing carbon nanotube tips will allow mechanical manipulation of many molecular systems on various surfaces with sub-angstrom accuracy. One particularly intriguing possibility along this line is to utilize a carbon nanotube SPM tip to engrave patterns on a silicon surface. It should be possible to create features a few nanometers across. These would be perhaps 100 times finer than the current state of the art in commercial semiconductor photolithography. Further, in contrast to approaches such as electron microscope lithography for which the speed of operation now appears to be an insuperable obstacle for industrial production, nanotube SPM-based lithography can be accelerated by utilizing an array with thousands of SPM tips simultaneously engraving different parts of a silicon surface. Also, nanotube SPM lithography could provide a practical means to explore various futuristic electronic device technology ideas, such as quantum cellular automata, which require exceedingly small feature sizes. Needless to say, if these ideas pan out, they could literally revolutionize computer device technology, paving the way for systems that are many times more powerful and more compact than any available today. For the near term, it should be noted that the semiconductor industry is a major market for SPM products. These are used to examine production equipment. High performance carbon nanotube tips should be of substantial value. NASA Ames is collaborating with Dr. Dai, now at Stanford, to develop these tips.

It is possible to store data on long chain molecules (for example, DNA) and it may be possible to read these data with carbon nanotube tipped SPMs. Existing DNA synthesis techniques can be used to write data. If the different DNA base pairs can be distinguished with a carbon nanotube tipped SPM, then the data can be read non-destructively (current techniques allow a destructive read). However, the difference between base pairs is not great. If the base pairs cannot be distinguished, techniques for attaching modified enzymes to specific base pair sequences could be used. Certain enzymes (DNA (cytosine-5) methyltransferases) attach themselves onto a specific sequence of base pairs with a covalent bond. The enzyme then performs its operation and breaks the bond. Smith modified the enzyme such that the initial covalent bond was formed but the subsequent operation was disrupted. The result is that DNA synthesized with the target base pair sequences at the desired location can force precise placement of the enzymes. The presence of an enzyme could represent 1 and its absence 0. Enzymes are sufficiently large that distinguishing their presence should be straightforward. If the DNA/enzyme approach proves impossible, a wide variety of other polymer systems could be examined.

Bauschlicher computationally studied storing data in a pattern of fluorine and hydrogen atoms on the (111) diamond surface. If write-once data could be stored this way, 10^{15} bytes/cm^2 is theoretically possible. By comparison, the new DVD write-once disks now coming on the market hold about 10^8 bytes/cm^2. Bauschlicher compared the interaction of different probe molecules with a one dimensional model of the diamond surface. This study found some molecules whose interaction energies with H and F are sufficiently different that the force differential should be detectable by an SPM. These studies were extended to include a two dimensional model of the diamond surface and two other systems besides F/H. Other surfaces, such as Si, and other probes, such as those including transition metal atoms, have also been investigated. Among the better probes was C_5H_5N (pyridine). Quantum calculations suggest that pyridine is stable when attached to C_{60} in the orientation necessary for sensing the difference between hydrogen and fluorine. Half of C_{60} can form the end cap of a (9,0) or (5,5) carbon nanotube, and carbon nanotubes have been attached to an SPM tip. Thus, it might be possible using today's technology to build a system to read the diamond memory surface. Avouris has shown that individual hydrogen atoms can be removed from a silicon surface. If this could be accomplished in a gas that donates fluorine to vacancies on a diamond surface, the data storage system could be built. Thummel computationally investigated methods for adding a fluorine at the radical sites where a hydrogen atom had been removed from a diamond surface.

As mentioned before, carbon nanotubes can be described as rolled up sheets of graphite. Different tubes can have different helical windings depending on how the graphite sheet is connected to itself. Theory suggests that single-walled carbon nanotubes can have metallic or semiconductor properties depending on the helical winding of the tube. Chico, Han, Menon, and others have computationally examined the properties of some of hypothetical devices that might be made by connecting tubes with different electrical properties. Such devices are only few nanometers across — 100 times smaller than current computer chip features. Several authors, including Tour have described methods to produce conjugated macromolecules of precise length and composition. This technique was used to produce molecular electronic devices in mole quantities. The resultant single molecular wires were tested experimentally and found to be conducting. The three and four terminal devices have been examined computationally and look promising. The features of these components are approximately 3 angstroms wide, about 750 times smaller than current silicon technology can produce.

Helical logic is a theoretical proposal for a future computing technology using the presence or absence of individual electrons (or holes) to encode 1s and 0s. The electrons are constrained to move along helical paths, driven by a rotating electric field in which the entire circuit is immersed. The electric field remains roughly orthogonal to the major axis of the helix and confines each charge carrier to a fraction of a turn of a single helical loop, moving it like water in an Archimedean screw. Each loop could in principle hold an independent carrier, permitting high information density. One computationally universal logic operation involves two helices, one of which splits into two "descendant" helices. At the point of divergence, differences in the electrostatic potential resulting from the presence or absence of a carrier in the adjacent helix controls the direction taken by a carrier in the splitting helix. The reverse of this sequence can be used to merge two initially distinct helical paths into a single outgoing helical path without forcing a dissipative transition. Because these operations are both logically and thermodynamically reversible, energy dissipation can be reduced to extremely low levels. ... It is important to note that this proposal permits a single electron to switch another single electron, and does not require that many electrons be used to switch one electron. The energy dissipated per logic operation can likely be reduced to less than 10^{-27} joules at a temperature of 1 Kelvin and a speed of 10 gigahertz, though further analysis is required to confirm this. Irreversible operations, when required, can be easily implemented and should have a dissipation approaching the fundamental limit of $\ln 2 \times kT$.

One study not conducted by Ames or partners is particularly worth mentioning since it places a loose lower bound on the computational capabilities of molecular nanotechnology. Drexler designed a number of computer components using small diamondoid rods with knobs that allow or prevent movement to accomplish computation. While this tiny mechanical Babbage Machine is probably not an optimal computational engine, its calculated performance for a desktop computer is 10^{18} MIPS — about a million times more powerful than the largest supercomputer that exists today. Note that with very fast computation energy use and heat dissipation become a severe problem. One approach to addressing this issue is reversible logic.

Drexler proposed a nanotechnology based on diamond and investigated its potential properties. In particular, he examined applications for materials with a strength similar to that of diamond (69 times strength/mass of titanium). This would require a very mature nanotechnology constructing systems by placing atoms on diamond surfaces one or a few at a time in parallel. Assuming diamondoid materials, McKendree predicted the performance of several existing single-stage-to-orbit (SSTO) vehicle designs. The predicted payload to dry mass ratio for these vehicles using titanium as a structural material varied from < 0 (the vehicle won't work) to 36 per cent, i.e., the vehicle weighs substantially more than the payload. With hypothetical diamondoid materials the ratios varied from 243 per cent to 653 per cent, i.e., the payload weighs far more than the vehicle. Using a very simple cost model ($1000 per vehicle kilogram) sometimes used in the aerospace industry, he estimated the cost per kilogram launched to low-Earth-orbit for diamondoid structured vehicles should be $153-412. This would meet NASA's 2020 launch to orbit cost goals. Estimated costs for titanium structured vehicles varied from $16,000-59,000/kg. Although this cost model is probably adequate for comparison, the absolute costs are suspect. Drexler used a more speculative methodology to estimate that a four passenger SSTO weighing three tons including fuel could be built using a mature nanotechnology. Using McKendree's cost model, such a vehicle would cost about $60,000 to purchase — the cost of today's high-end luxury automobiles. These studies assumed a fairly advanced nanotechnology capable of building diamondoid materials. In the nearer term, it may be possible to develop excellent structural materials using carbon nanotubes. Carbon nanotubes have a Young's modulus of approximately one terapascal— comparable to diamond.

Issacs and Pearson proposed a space elevator—a cable extending from the Earth's surface into space with a center of mass at geosynchronous altitude. If such a system could be built, it should be mechanically stable and vehicles could ascend and descend along the cable at almost any reasonable speed using electric power (actually generating power on the way down). The first incredibly difficult problem with building a space elevator is strength of materials. Maximum stress is at geosynchronous altitude so the cable must be thickest there and taper exponentially as it approaches Earth. Any potential material may be characterized by the taper factor—the ratio between the cable's radius at geosynchronous altitude and at the Earth's surface. For steel the taper factor is tens of thousands—clearly impossible. For diamond, the taper factor is 21.9 including a safety factor. Diamond is, however, brittle. Carbon nanotubes have a strength in tension similar to diamond, but bundles of these nanometer-scale radius tubes shouldn't propagate cracks nearly as well as the diamond tetrahedral lattice. Thus, if the considerable problems of developing a molecular nanotechnology capable of making nearly perfect carbon nanotube systems approximately 70,000 kilometers long can be overcome, the first serious problem of a transportation system capable of truly large scale transfers of mass to orbit can be solved. The next immense problem with space elevators is safety—how to avoid dropping thousands of kilometers of cable on Earth if the cable breaks. Active materials may help by monitoring and repairing small flaws in the cable and/or detecting a major failure and disassembling the cable into small elements.

Drexler calculates that lightsails made of 20 nm aluminum in tension should achieve an outward acceleration of ~14 km/s per day at Earth orbit with no payload and minimal structural overhead. For comparison, the delta V from low Earth to geosynchronous orbit is 3.8 km/s. Lightsails generate thrust by reflecting sunlight. Tension is achieved by rotating the sail. The direction of thrust is normal to the sail and away from the Sun. By directing thrust along or against the velocity vector, orbits can be lowered or raised. This form of transportation requires no reaction mass and generates thrust continuously, although the instantaneous acceleration is small so sails cannot operate in an atmosphere and must be large for even moderate payloads.

Today, the smallest feature size in production systems is about 250 nanometers — the smallest feature size in computer chips. Since atoms are an angstrom or so across and carbon nanotubes have a diameter as small as 0.7 nanometers, atomically precise molecular machines can be smaller than current MEMS devices by two to three orders of magnitude in each dimension, or six to nine orders of magnitude smaller in volume (and mass). For example, the size of the kinesin motor, which transports material in cells, is 12 nm. Han computationally demonstrated that molecular gears fashioned from single-walled carbon nanotubes with benzyne teeth should operate well at 50-100 gigahertz. These gears are about two nanometers across. Han computationally demonstrated cooling the gears with an inert atmosphere. Srivastava simulated powering the gears using alternating electric fields generated by a single simulated laser. In this case, charges were added to opposite sides of the tube to form a dipole. To make active materials, a material might be filled with nano-scale sensors, computers, and actuators so the material can probe its environment, compute a response, and act. Although this document is concerned with relatively simple artificial systems, living tissue may be thought of as an active material. Living tissue is filled with protein machines which gives living tissue properties (adaptability, growth, self-repair, etc.) unimaginable in conventional materials.

Active materials can theoretically be made entirely of machines. These are sometimes called swarms since they consist of large numbers of identical simple machines that grasp and release each other and exchange power and information to achieve complex goals. Swarms change shape and exert force on their environment under software control. Although some physical prototypes have been built, at least one patent issued, and many simulations run, swarm potential capabilities are not well analyzed or understood. We briefly discuss some concepts here. Michael proposes brick-shaped machines of various sizes that slide past each other to assume a variety of shapes. He has generated a large number of videos showing computer simulations of simple motions. Although his web site contains rather extravagant claims, this work has received a U.K. patent.

Yim built a small swarm with macroscopic (size in inches) components called polypod, built a simulator of polypod, and programmed it to move in various ways to study locomotion. There are two brick shaped components in polypod, one of which has two prismatic joints linked by a revolute joint. The second component is a cubic connector with no mechanical motion. Polypod is programmed by tables for each member of the swarm. Each member is programmed to move at various speeds in each degree of freedom for certain amounts of time. The swarm components are implicitly synchronized so there is no clock signal. Hall proposes a swarm with 10 micron dodecahedral components each with 12 arms that can move in and out, rotate a little, and grab and release each other. This concept is called the "utility fog." Hall estimates that the utility fog would have a density of 0.2, tensile strength of 1000 psi in action and 100,000 psi in a passive mode, and have a maximum shear rate of 100 km/second/meter. Bishop proposes a swarm consisting of 100 nanometer brick-shaped components that slide past each other to change shape. Globus proposes a swarm with two kinds of components — edges and nodes. The terms "node" and "edge" are chosen to correspond to those in graph theory. The roughly spherical nodes are capable of attaching to five edges (for a tetrahedral geometry with one free edge per node) and rotating each edge in pitch and yaw. The rod-like edges are capable of changing length, rotating around their long axis, and attaching/detaching to/from nodes. Component design, power distribution and control software are significant challenges for swarm development. Consider that with 10 micron components a cubic meter of swarm would contain about 10^{15} devices, each with an internal computer communicating with its neighbors to accomplish a global task.

NASA's mission is divided into five enterprises: Mission to Planet Earth, Aeronautics, Human Exploration and Development of Space, Space Science, and Space Technology. We will examine some potential nanotechnology applications in each area.

The Earth Observing System (EOS) will use satellites and other systems to gather data on the Earth's environment. The EOS data system will need to process and archive >terabyte per day for the indefinite future. Simply storing this quantity of data is a significant challenge—each day's data would fill about 1,000 DVD disks. With projected nanocomputer processing speeds of 10^{18} MIPS, a million calculations on each byte of one day's data would take one second on the desktop.

Given a mature nanotechnology, it should be possible to build sensors in balloon-borne systems approximately the size of bacteria. With replication based manufacturing, these should be quite inexpensive. If the serious communication and control problems can be solved, one can imagine spreading billions of tiny lighter-than-air vehicles into the atmosphere to measure wind currents and atmospheric composition. A similar approach might be taken in the oceans—note that the oceans are full of floating microscopic living organisms that can sense and react to their environment. Smart dust might sense the environment, note the location via a GPS-like system, and store that information until close enough to a data-collection point to transfer the data to the outside world.

The strength of materials and computational capabilities previously discussed for space transportation should also allow much more advanced aircraft. Stronger, lighter materials can obviously make aircraft with greater lift and range. More powerful computers are invaluable in the design stage and of great utility in advanced avionics.

MEMS technology has been used to replace traditional large control structures on aircraft with large numbers of small MEMS controlled surfaces. This control system was used to operate a model airplane in a windtunnel. Nanotechnology should allow even finer control—finer control than exhibited by birds, some of which can hover in a light breeze with very little wing motion. Nanotechnology should also enable extremely small aircraft.

A reasonably advanced nanotechnology should be able to make simple atomically precise materials under software control. If the control is at the atomic level, then the full range of shapes possible with a given material should be achievable. Aircraft construction requires complex shapes to accommodate aerodynamic requirements. With molecular nanotechnology, strong complex-shaped components might be manufactured by general purpose machines under software control.

The aeronautics mission is responsible for launch vehicle development. Payload handling is an important function. Very efficient payload handling might be accomplished by a very advanced swarm. The sequence begins by placing each payload on a single large swarm located next to the shuttle orbiter. The swarm forms itself around the payloads and then moves them into the payload bay, arranging the payloads to optimize the center of gravity and other considerations. The swarm holds the payload in place during launch and may even damp out some launch vibrations. On orbit, satellites can be launched from the payload bay by having the swarm give them a gentle push. The swarm can then be left in orbit, perhaps at a space station, and used for orbital operations. This scenario requires a very advanced swarm that can operate in an atmosphere and on orbit in a vacuum. Besides the many and obvious difficulties of developing a swarm for a single environment, this provides additional challenges. Note that a simpler swarm might be used for aircraft payload handling.

Aerospace vehicles often require complex checkout procedures to insure safety and reliability. This is particularly true of reusable launch vehicles. A very advanced swarm with some special purpose appendages might be placed on a vehicle. It might then spread out over the vehicle and into all crevices to examine the state of the vehicle in great detail. Nanotechnology-enabled Earth-to-orbit transportation has the greatest potential to revolutionize human access to space by dropping the current $10,000 per pound cost of launch, but this was discussed above. Other less dramatic technologies include:

Space structures with a long design life (such as space station modules) need high-reliability materials that do not degrade. Active materials might help. The machines monitor structural integrity at the sub-micrometer scale. When a portion of the material becomes defective, it could be disassembled and then correctly reassembled. It should be noted that bone works somewhat along these lines. It is constantly being removed and added by specialized cells.

To effect timely repairs, space stations require a large store of spare parts and tools that are rarely used. A mature nanotechnology might create a "matter compiler," a machine that converts raw materials into a wide variety of products under software control. Contemporary examples of very limited matter compilers are numerically controlled machines and polypeptide sequencers. With a substantially more capable nanotechnology-based matter compiler, a space station crew could simply make spare parts and tools as needed. The programs could be stored on-board or on the ground. New tools invented on Earth could be transferred as software to the station for manufacture. Once used, unneeded tools and broken parts could be ionized in a solar furnace, transferred using controlled magnetic fields, and the constituent atoms stored for later manufacture into new products.

An advanced nanotechnology might be able to build filters that dynamically modify themselves to attract the contaminant molecules detected by the air and water quality sensors. Once attached to the filter, the filter could in principle move the offending molecules to a molecular laboratory for modifications to useful or at least inert products. A swarm might implement such an active filter if it was able to dynamically manufacture proteins that could bind contaminant molecules. The protein and bound contaminant might then be manipulated by the swarm for transportation. With a sufficiently advanced nanotechnology it might even be possible to directly generate food by non-biological means. Then agriculture waste in a self-sufficient space colony could be converted directly to useful nutrition. Making this food attractive will be a major challenge. Sleeping crew members in the shuttle experience considerable pain and sleep disruption when the reaction control system fires and they collide with the cabin walls. If crew members were connected to the walls by a swarm, the swarm could absorb most or all of the force before the crew member struck the wall. The swarm could then gradually return the crew member to center (without the oscillations associated with bungee cords) in preparation for the next firing. For resupply, spacecraft docking is a frequent necessity in space station operations. When two spacecraft are within a few meters of each other, a swarm could extend from each, meet in the middle, and form a stable connection before gradually drawing the spacecraft together. A swarm could support space-suited astronauts in simulated partial-g environments by holding them up appropriately. The swarm moves in response to the astronaut's motion providing the appropriate simulation of partial or 0 gravities. Tools and other objects are also manipulated by the swarm to simulate non-standard gravity.

Active nanotechnology materials (see active materials) might enable construction of a skin-tight space suit covering the entire body except the head (which is in a more conventional helmet). The material senses the astronaut's motions and changes shape to accommodate it. This should eliminate or substantially reduce the limitations current systems place on astronaut range of motion.

In situ resource utilization is undoubtedly necessary for large scale colonization of the solar system. Asteroids are particularly promising for orbital use since many are in near Earth orbits. Moving asteroids into low Earth orbit for utilization poses a safety problem should the asteroid get out of control and enter the atmosphere. Very small asteroids can cause significant destruction. The 1908 Tunguska explosion, which was calculated to be a 60 meter diameter stony asteroid, leveled 2,200 km^2 of forest. Hills calculated that 4 meter diameter iron asteroids are near the threshold for ground damage. Both these calculations assumed high collision speeds. At a density of 7.7 g/cm^3 Babadzhanov, a 3 meter diameter asteroid should have a mass of about 110 tons. Rabinowitz estimates that there are about one billion ten meter diameter near Earth asteroids and there should be far more smaller objects. For colonization applications one would ideally provide the same radiation protection available on Earth. Each square meter on Earth is protected by about 10 tons of atmosphere. Therefore, structures orbiting below the van Allen belts would like 10 tons/meter^2 surface area shielding mass. This would dominate the mass requirements of any system and require one small asteroid for each 11 meter^2 of colony exterior surface area. A 10,000 person cylindrical space colony such as Lewis One with a diameter of almost 500 meters and a length of nearly 2000 meters would require a minimum of about 90,000 retrieval missions to provide the shielding mass. The large number of missions required suggests that a fully automated, replicating nanotechnology may be essential to build large low Earth orbit colonies from small asteroids. A nanotechnology swarm along with an atomically precise lightsail is a promising small asteroid retrieval system. Lightsail propulsion insures that no mass will be lost as reaction mass. The swarm can control the lightsail by shifting mass. When a target asteroid is found, the swarm spreads out over the surface to form a bag. The interface to the sail must be active to account for the rotation of the asteroid—which is unlikely to have an axis-of-rotation in the proper direction to apply thrust for the return to Earth orbit. The active interface is simply swarm elements that transfer between each other to allow the sail to stay in the proper orientation. Of course, there are many other possibilities for nanotechnology based retrieval vehicles.

Extraterrestrial materials brought into orbit could be fed into a high-temperature solar furnace and partially ionized. Magnetic fields might then be used to separate the nuclei. These are fed in appropriate quantities to a matter-compiler to build the products desired. Several authors, including Freitas have speculated that a sufficiently advanced nanotechnology could examine and repair cells at the molecular level. Should this capability become available—presumably driven by terrestrial applications—the small size and advanced capabilities of such systems could be of great utility on long duration space flights and on self-sufficient colonies.

Self-replicating systems permit efforts of great scope to be pursued economically. Adjusting the environment on another planet to suit the tastes of humans is one such undertaking. Heating and cooling can be achieved by (among many other methods) using space-based mirrors. Chemical modifications of the planetary surface and atmosphere can be achieved in relatively short periods by the use of self-replicating systems that absorb sunlight and raw materials, and convert them into the desired products. Much as plants changed the environment of the earth to what we see today, so self-replicating molecular manufacturing systems might more rapidly convert the environments of other planets.

As interstellar trips might last many years, the ability to conserve supplies by maintaining some crew members in a suspended state would be useful. An extremely advanced nanotechnology might use molecular manipulations of each cell to provide (a) better methods of slowing or suspending the metabolic activity of crew members and (b) better methods of restoring metabolic activity to a normal state when the destination is reached. Molecular manufacturing should enable the creation of very precise mirrors. Unlike lightsail applications, telescope mirrors require a very precise and somewhat complicated shape. A swarm with special purpose appendages capable of bonding to the mirror might be able to achieve and maintain the desired shape. A very advanced nanotechnology would be capable of imaging and then removing the surface atoms of an extra-terrestrial sample. By removing successive surface layers the location of each atom in the sample might be recorded, destroying the sample in the process. This data could then be sent to Earth. Besides requiring a very advanced nanotechnology, there is a more fundamental—but not necessarily fatal—problem: as the outside layer of atoms is removed the next layer may rearrange itself so the sample is not necessarily perfectly recorded.

Solar Power—Low Earth orbit spacecraft generally depend on solar cells and batteries for power. For energy collection, molecular manufacturing can be used to make solar photovoltaic cells at least as efficient as those made in the laboratory today. Efficiencies can therefore be > 30 per cent. In space applications, a reflective optical concentrator need consist of little more than a curved aluminum shell < 100 nanometers thick (photovoltaic cells operate with higher efficiency at high optical power densities). A metal fin with a thickness of 100 nanometers and a conduction path length of 100 microns can radiate thermal energy at a power density as high as 1000 W/m^2 with a temperature differential from base to tip of < 1 K. Accordingly, solar collectors can consist of arrays of photovoltaic cells several microns in thickness and diameter, each at the focus of a mirror of ~100 micron diameter, the back surface of which serves as a ~100 micron diameter radiator. If the mean thickness of this system is ~1 micron, the mass is ~10^{-3} kg/m^2 and the power per unit mass, at Earth's distance from the Sun, where the solar constant is ~1.4 kW/m^2, is > 10^5 W/kg." By comparison, the U.S. built Photovoltaic Panel Module solar cells currently used on the Mir Space Station and planned for use on the International Space Station generate about 118 W/kg.

Fuel Cells—A critical component in hydrogen/oxygen fuel cells is the PEM (Proton Exchange Membrane). This membrane must (a) permit the passage of protons while (b) blocking everything else. Present membranes do a rather poor job. One group at Ames is designing and computationally testing PEMs to study possible energy mechanisms in early life. While these studies are not meant to design optimal membranes for fuel cell use, the basic knowledge and approach may be of value. Another proposal is to design a diamond membrane a few nanometers thick with "proton pores." The pores might be lined with fluorine, oxygen and nitrogen to create a region with a high proton affinity. In addition, a positionally controlled platinum might be held at the mouth of the pore to verify that H_2 can be catalytically split into H+ and e–, and that the barrier for migration of the H+ into the pore is modest in size. Nanotechnology must provide precise control over the manufacturing process of the diamondoid PEM since the pores must be made very precisely.

Studies of H_2 absorption and packing in carbon nanotubes and nanoropes are in progress at NASA Ames and elsewhere. Nanotubes provide large pore sizes and nanoropes have different pore sizes depending on interstitial and other locations. Dillon estimated that the single walled nanotubes in their sample contained 5 to 10 per cent by weight of H_2. The nanotubes were about 0.1 to 0.2 per cent by weight of the total sample. Computational studies at Ames suggest that to store 7-10 per cent H_2 in single walled nanotubes at room temperature the H_2s must be stored inside the tubes, not merely adsorbed on the walls. This work suggests that carbon nanotubes might be developed into an excellent H_2 storage medium within 3-5 years. Calculations with oxygen suggest that a diamondoid sphere ~0.1 microns in diameter should easily hold oxygen at ~1,000 atmospheres. While higher pressures are feasible, they offer declining returns. At higher pressures, the pressure-volume relationship becomes severely non-linear and the density approaches a limiting value. Other gases might also be stored if diamondoid spheres can be built, but the analysis has not been done. High strength light-weight materials will allow greater efficiency of energy storage as angular momentum.

Many kinds of ultraminiature electromechanical devices have utility on a miniaturized space craft. It has been shown that manipulating carbon nanotubes changes their electrical properties. This might be exploited to build nanometer scale strain devices. This may be achievable within 3-5 years, and simulations along these lines are in progress. Similar results have been achieved experimentally with C_{60}. The electrical properties of a C_{60} molecule were changed by applying pressure to the molecule with an SPM tip. Smaller, lighter spacecraft are cheaper to launch (current costs are about

$10,000/lb) and generally cheaper to build. Diamondoid structural materials can radically reduce structural mass, miniaturized electronics can shrink the avionics and reduce power consumption, and atomically precise materials and components should shrink most other subsystems. Thermal protection is crucial for atmospheric reentry and other tasks. The carbon nanotubes under investigation at NASA Ames and elsewhere may play a significant role. Most production processes for carbon nanotubes create a tangled mat of nanotubes that has a very low mass-to-volume ratio. Like graphite, the tubes should withstand high temperatures but the tangled mat should prevent them from ablating. This may lead to high temperature applications.

Many of the applications discussed here are speculative to say the least. However, they do not appear to violate the laws of physics. Something similar to these applications at these performance levels should be feasible if we can gain complete control of the three-dimensional structure of materials, processes and devices at the atomic scale. How to gain such control is a major, unresolved issue. However, it is clear that computation will play a major role regardless of which approach —positional control with replication, self-assembly, or some other means —is ultimately successful. Computation has already played a major role in many advances in chemistry, SPM manipulation, and biochemistry. As we design and fabricate more complex atomically precise structures, modeling and computer aided design will inevitably play a critical role. Not only is computation critical to all paths to nanotechnology, but for the most part the same or similar computational chemistry software and expertise supports all roads to molecular nanotechnology. Thus, even if NASA's computational molecular nanotechnology efforts should pursue an unproductive path, the expertise and capabilities can be quickly refocused on more promising avenues as they become apparent. As nanotechnology progresses we may expect applications to become feasible at a slowly increasing rate. However, if and when a general purpose programmable assembler/replicator can be built and operated, we may expect an explosion of applications. From this point, building new devices will become a matter of developing the software to instruct the assembler/replicators. Development of a practical swarm is another potential turning point. Once an operational swarm that can grow and divide has been built, a large number of applications become software projects. It is also important to note that the software for swarms and assembler/replicators can be developed using simulators — even before operational devices are available. Nanotechnology advocates and detractors are often preoccupied with the question "When?" There are three interrelated answers to this question:

1. Nobody knows. There are far too many variables and unknowns. Beware of those who have excessive confidence in any date.
2. The time-to-nanotechnology will be measured in decades, not years. While a few applications will become feasible in the next few years, programmable assembler/replicators and swarms will be extremely difficult to develop.

3. The time-to-nanotechnology is very sensitive to the level of effort expended. Resources allocated to developing nanotechnology are likely to be richly rewarded, particularly in the long term.

REFERENCES

Andres, R.P., S. Datta, D.B. Janes, C.P. Kubiak, and R. Reifenberger. 1998. The design, fabrication and electronic properties of self-assembled molecular nanostructures. In *The handbook of nanostructured materials and nanotechnology*, ed. H.S. Nalwa. San Diego: Academic Press.

Asada, M., Y. Mayamoto, and Y. Suematsu. 1986. *IEEE Journ. Quantum Electronics* QE- 22(9): 1915-1921.

Attard, G.S., et al. 1997. *Science* 278:838. Averback, R.S., J. Bernholc, and D.L. Nelson. 1991. *MRS Symposium Proceedings* Vol. 206.

Axelbaum, R.L. 1997. Developments in sodium/halide flame technology for the synthesis of unagglomerated non-oxide nanoparticles. In *Proc. of the Joint NSF-NIST Conference on Nanoparticles: Synthesis, Processing into Functional Nanostructures and Characterization* (May 12-13, Arlington, VA).

B. W. Matthews, H. Nicholson, and W. J. Becktel, "Enhanced protein thermostability from site-directed mutations that decrease the entropy of unfolding." *Proc. Nat. Acad. Sci., 84:* 6663-6667 (1987), and included references.

Baibich, M..N., J.M. Broto, A. Fert, F. Nguyen Van Dau, F. Petroff, P. Etienne, G. Creuzet, A. Friederich, and J. Chazelas. 1998. *Phys. Rev. Lett.* 61:2472.

Baird, D.: 2004, *Thing Knowledge: A Philosophy of Scientific Instruments*, University of California Press, Berkeley.

Baker, R.T.K. Synthesis, properties and applications of graphite nanofibers. In *R&D status and trends*, ed. Siegel et al.

Beck, J.S., J.C. Vartuli, W.J. Roth, M.E. Leonowicz, C.T. Kresge, K.D. Schmitt, C.T.-W. Chu, D.H. Olsen, E.W. Shepard, S.B. McCullen, J.B. Higgins, and J.L. Schlenker. 1992. *J. Am. Chem. Soc.* 114:10834.

Becker, M.F., J.R. Brock, H. Cai, N. Chaudhary, D. Henneke, L. Hilsz, J.W. Keto, J. Lee, W.T. Nichols, and H.D. Glicksman. 1997. Nanoparticles generated by laser ablation. In *Proc. of the Joint NSF-NIST Conf. on Nanoparticles*.

Berger, R., C. Gerber, H.P. Lang, and J.K. Gimzewski. 1996. Micromechanics: a toolbox for femtoscale science: "Towards a laboratory on a tip." *Microelectronic Engineering*, 35:373-9. (International Conference on Micro- and Nanofabrication, Glasgow, U.K., 22- 25 Sept. 1996).

Berkowitz, A.E., J.R. Mitchell, M.J. Carey, A.P. Young, S. Zhang, F.E. Spada, F.T. Parker, A. Hutten, and G. Thomas. 1992. *Phys. Rev. Lett.* 68:3745.

Berndt, C.C., J. Karthikeyan, T. Chraska, and A.H. King. 1997. Plasma spray synthesis of nanozirconia powder. In *Proc. of the Joint NSF-NIST Conf. on Nanoparticles*.

Bimberg, D., N. Kirstaedter, N.N. Ledentsov, Zh.I. Alferov, P.S. Kopev, and V.M. Ustinov. 1998. InGaAs-GaAs quantum-dot lasers. *IEEE J. of Selected Topics in Quant. Electronics* 3:196-205.

Bowes, C.L., A. Malek, and G.A. Ozin. 1996. *Chem. Vap. Deposition* 2:97.

Braun, P.V., P. Osenar, and S.I. Stupp. 1996. *Nature*. 368: 2. Brinker, C.J. 1996. *Curr. Opin. Solid State Mater. Sci.* 1:798. Chen, C-Y., S.L. Burkett, H.-X. Li, and M.E. Davis. 1993. *Microporous Mater.* 2:27.

Brotzman, R. 1998. Nanoparticle Dispersions. In *R&D status and trends*, ed. Siegel et al. Bu, X., P. Feng, and G.D. Stucky. 1998. Large-cage zeolite structures with multidimensional 12-ring channels. *Science* 278:2080-2085.

Brotzman, R. 1998. Nanoparticle dispersions. In *R&D status and trends*, ed. Siegel et al. Brus, L.E. 1996. Theoretical metastability of semiconductor crystallites in high pressure phases with applications to beta tin structures of silicon. *J. Am. Chem. Soc.* 118:4834-38.

Brus, L. 1996. Semiconductor colloids: Individual nanocrystals, opals and porous silicon. *Current Opinion in Colloid & Interface Science* 1:197-201. *Chemical Engineering News*. 1997. Particulate matter health studies to be reanalyzed (August 18):33.

Calcote, H.F., and D.G. Keil. 1997. Combustion synthesis of silicon carbide powder. In *Proc. of the Joint NSF-NIST Conf. on Nanoparticles*.

Canguilhem, G.: 1952, *Machine et organisme' in La connaissance de la vie*, Hachette, Paris [quoted from the fourth edition Vrin, Paris, 1971].

Carsley, J.E., R. Shaik, W.W. Milligan, and E.C. Aifantis. 1997. In *Chemistry and physics of nanostructures and related non-equilibrium materials*. ed. E. Ma, B. Fultz, R. Shull, J. Morral, and P. Nash. Warrendale, PA: TMS.

Chance, B., Mueller, P., DeVault, D. & Powers, L. (1980) *Phys. Today* 33 (10), 32-38.

Chang, K. "Smaller Computer Chips Built Using DNA as Template." New York Times, November 21, 2003.

Chantrell. 1994. *J. Appl. Phys.* 76:6811. Erb, U., G. Palumbo, R. Zugic, and K.T. Aust. 1996. In *Processing and properties of nanocrystalline materials*, ed. C. Suryanarayana, J. Singh, and F.H. Froes. Warrendale, PA: TMS. Gertsman, V.Y., M. Hoffman, H. Gleiter and R. Birringer. 1994. *Acta Metall. Mater.* 42:3539-3544.

Chopra, N.G., R.J. Luyken, K. Cherrey, H. Crespi, M.L. Cohen, S.G. Louie, and A. Zettl. 1995. *Science* 269:966.

Clasen, R. 1990. *Int. Journal of Glass and Science Technology* 63: 291. Crandall, B.C. and J. Lewis, (eds.). 1992. *Nanotechnology: Research and perspectives*. Cambridge: MIT Press.

Czekai, D., et al. 1994. *Use of smaller milling media to prepare nanoparticulate dispersions*. U.S. Patent application. Docket 69802(02) Filed 2/25/94.

Dai, H., E.W. Wong, Y.Z. Lu, S. Fan, and C.M. Leiber. 1995. *Nature* 375:769. deHeer, W.A., W.S. Bacsa, A. Chatelain, T. Gerfin, R. Humphrey-Baker, L. Forro, and D. Ugarte. 1995. *Science* 268:845.

Dresselhaus, M.S., G. Dresselhaus, and P. Eklund. 1996. *Science of fullerenes and carbon nanotubes*. San Diego: Academic Press. Gleiter, H. 1989. *Prog. Mater. Sci.* 33:223.

Drexler K.E.: 1995, 'Introduction to nanotechnology', in: Krummenacker, M. & Lewis, J. (eds.), *Prospects in Nanotechology. Proceedings of the 1st general conference on nanotechnology: developments, applications, and opportunities, November 11-14, 1992, Palo-Alto*, John Wiley & Sons, New York, pp. 1-20.

Drexler, E., Phoenix, C.: 2004, 'Self-Replication in Nanotechnology: Feasible, Potentially Safe, and Unnecessary' (unpublished paper, in preparation).

Drexler, E.: 1986, *Engines of Creation: The Coming Era of Nanotechnology*, Anchor Books, New York (expanded edition with a new afterword, 1990).

Drexler, E.: 1992, *Nanosystems: Molecular Machinery, Manufacturing, and Computation*, John Wiley & Sons, New York.

Drexler, K.E.: 1981, 'Molecular engineering: An approach to the development of general capabilities for molecular manipulation', *Proceedings of the National Academy of Sciences*, 78, no. 9, chemistry section, 5275-78.

Drexler, K.E.: 1992, *Nanosystems. Molecular machinery, manufacturing and computation*, John Wiley & Sons, New York.

Drexler, K.E.: 2001, 'Machine-Phase nanotechnology', *Scientific American*, (Sept.), 66-67.

Drexler, K.E.: 2003a, 'Open Letter', *Chemical & Engineering News*, 81, no. 48, 37-42.

Drexler, K.E.: 2003b, 'An open letter to Richard Smalley', April 16, [published on KurzweilAI.net].

Dupuy, J.P. "Common Knowledge, Common Sense." Theory and Decision 27 (1989): 37-62.

Dupuy, J.P. "Two Temporalities, Two Rationalities: A New Look at Newcomb's Paradox." in Bourgine, P.&Walliser, B. (eds.), Economics and Cognitive Science. New York: Pergamon. 1992, 191-220.

Dupuy, J.P.: 2000, *The Mechanization of the Mind*, Princeton University Press, Princeton N.J.

Ellsberg, D. "Risk, Ambiguity and the Savage Axioms." Quarterly Journal of Economics 75 (1961): 643-669.

Entingh, D., Dunn, A., Glassman, E., Wilson, J. E., Hogan, E. & Damstra, T. (1975) in *Handbook of Psychobiology*, eds. Gazzinga, M. S. & Blakemore, C. (Academic, New York), pp. 201-238.

Estermann, M., L.B. McCusker, C. Baerlocher, A. Merrouche, and H. Kessler. 1991. *Nature* 331:698.

Esumi, A., A. Suzuki, N. Aihara, K. Uswi, and K. Torigoe. 1998. *Langmuir*, 14:3157.

Evans, D.F.; Wennerstrom, H.: 1999, *The Colloidal Domain: Where Physics, Chemistry, Biology and Technology Meet*, John Wiley & Sons, New York.

Faraday, M.: 1857, 'Experimental relations of gold (and other metals) to light', *Philosophical Transactions of the Royal Society London*, 147, 145-181.

Fennema, O. R. (1973) in *Low-Temperature Preservation of Foods and Living Matter*, eds. Fennema, O. R., Powrie, W. D. & Marth, E. H. (Dekker, New York), pp. 436-475.

Feyerabend, P.: 1981, *Realism, Rationalism and Scientific Method* (Philosophical Papers, volume 1), Cambridge University Press, Cambridge.

Feynman R. "There's Plenty of Room At the Bottom." Talk given on at the annual meeting of the American Physical Society at the California Institute of Technology, 1959.

Feynman, R.: 1960, 'There's Plenty of Room at the Bottom', *Engineering and Science*, 23, 22-36.

Fleming, D. "The Economics of Taking Care: An Evaluation of the Precautionary Principle." in Freestone, D.&Hey, E. (eds.), The Precautionary Principle and International Law. La Haye: Kluwer Law International, 1996.

Foerster, H.v.: 1962, 'Bio-Logic', in: E.E. Bernard & M.R. Kare (eds.), *Biological prototypes and synthetic systems*, Plenum Press, New York, vol. 1., pp. 1-12

Friedlander, S.K. 1993. Controlled synthesis of nanosized particles by aerosol processes. *Aerosol Sci. Technol.* 19:527.

Friedlander, S.K. 1998. Synthesis of nanoparticles and their agglomerates: Aerosol reactors. In *R&D status and trends*, ed. Siegel et al.

Friedlander, S.K., H.D. Jang, and K.H. Ryu. 1998. *Appl. Phy. Lett.* 72(2):173.

Gleiter, H. 1989. *Prog. Mater. Sci.* 33:223. Goddard, W.A. 1998. Nanoscale theory and simulation. In *R&D status and trends*, ed. Siegel et al.

Gleiter, H. 1990. *Progress in Materials Science* 33:4. Guinier, A. 1938. *Nature* 142:569; Preston, G.D. 1938. *Nature* 142:570.

Goddard, W.A. 1998. Nanoscale theory and simulation. In *R&D status and trends*, ed. R. Siegel et al.

Godsell, D.: 2003, *Living Machinery. Bionanotechnology: Lessons from Nature*, Wiley-Liss, New-York.

Guarnieri, F., M. Fliss, and C. Bancroft. 1996. Making DNA Add. *Science* 273:220-223. Hanes, J., J.L. Cleland, and R. Langer. 1997.

Günther, B., A. Baalmann, and H. Weiss. 1990. *Mater. Res. Soc. Symp. Proc.* 195:611-615.

Gutte, B., Dannigen, M. & Wittschieber, E. (1979) *Nature (London)* 281, 650-655.

Hadjipanayis, C.G. 1998. Nanostructured magnetic materials. In *R&D Status and Trends*, ed. Siegel et al.

Han, W., S. Fam, O. Li, and Y. Hu. 1998. Synthesis of gallium nitride nanorods through a carbon nanotube-confined reaction. *Science* 277:1287-1289.

Hansson, S.O. "The False Promises of Risk Analysis." Ratio 6 (1993): 16-26.

Held, R., T. Heinzel, A.P. Studerus, K. Ensslin, and M. Holland. 1997. Semiconductor quantum point contact fabricated by lithography with an atomic force microscope. *Appl. Phys. Lett.* 71:2689-91.

Hellemans, A. 1998. X-rays find new ways to shine. *Science* 277:1214-15.

Hietpas, P.B., S.D. Gilman, R.A. Lee, M.R. Wood, N. Winograd, and A.G. Ewing. 1996. Development of votammetric methods, capillary electrophoresis and tof sims imaging for constituent analysis of single cells. In *Nanofabrication and biosystems*, ed.

Higgins, R.J. 1997. An economical process for manufacturing of nano-sized powders based on microemulsion-mediated synthesis. In *Proc. of the Joint NSF-NIST Conf. on Nanoparticles.*

Hiruma, K., M. Yazawa, T. Katsoyama, K. Ogawa, K. Haraguchi, M. Koguchi, and H. Kakibayashi. 1995. *J. Appl. Phys.* 77(2):476.

Hoch et al. Ho, S.V., P.W. Sheridan, and E. Krupetsky. 1996. Supported polymeric liquid membranes for removing organics from aqueous solutions.

Hoch, H.C., L.W. Jelinski, and H.G. Craighead, eds. 1996. *Nanofabrication and biosystems.* New York: Cambridge University Press.

Howland, H.: 1962, Structural, hydraulic and economic aspects of leaf venation and shape ', in: E.E. Bernard & M.R. Kare (eds.), *Biological prototypes and synthetic systems*, Plenum Press, New York, vol. 1, pp. 183-192 .

Hoyningen-Huene, P.: 1993, *Reconstructing Scientific Revolutions: Thomas S. Kuhn's Philosophy of Science* (trans. by A. Levin), University of Chicago Press, Chicago.

Huang, J.Y., Y.K. Wu, and H.Q. Ye. 1996. *Acta Mater*. 44:1211. Inoue, A. 1997. Private communication. Inturi, R.B., and Z. Szklavska-Smialowska. 1992. *Corrosion* 48:398. Karch, J., R. Birringer, and H. Gleiter. 1987. *Nature* 330:556.

Hubbell, J.A., and R. Langer. 1995. Tissue engineering. *Chem. Eng. News* (March 13): 42- 54.

Huo, Q., D.I. Margolese, U. Ciesla, P. Feng, T.E. Gier, P. Sieger, R. Leon, P.M. Petroff, F. Schuth, and G.D. Stucky. 1994. *Nature* 368:317.

Imae, Y., and T. Atsumi. 1989. T. Na+-driven bacterial flagellar motors: A mini-review. *J. Bioenergetics and Biomembranes* 21:705-716.

Itakura, K. & Riggs, A. D. (1980) *Science* 209, 1401-1405.

Jaworek, T., D. Deher, G. Wegner, R.H. Wieringa, and A.J. Schouten. 1998. Electromechanical properties of an ultrathin layer of directionally aligned helical polypeptides. *Science* 279:57-60.

Jena, P., S.N. Khanna, and B.K. Rao. 1996. In *Science and technology of atomically engineered materials*, ed. P. Jena. River Edge, NJ: World Scientific.

Jonas, H. The Imperative of Responsibility. In Search of an Ethics for the Technological Age. Chicago: University of Chicago Press, 1985.

Jounet, C., W.K. Maser, P. Bernier, A. Loiseau, M. Lamy de la Chapelle, S. Lefrant, P. Deniard, R. Lee, and J.E. Fischer. 1997. *Nature* 388:756.

Junno, T., S.-B. Carlsson, H. Xu, L. Montelius, and L. Samuelson. 1998. Fabrication of quantum devices by angstrom-level manipulation of nanoparticles with an atomic force microscope. *Appl. Phys. Lett.* 72:548-550.

K. E. Drexler, "Molecular engineering: An approach to the development of general capabilities for molecular manipulation." *Proc. Nat. Acad. Sci.*, 78: 5275-5258 (1981).

K. E. Drexler, *Engines of Creation*, Anchor/Doubleday (New York, 1986).

Karak, N., and S. Maiti. 1997. Dendritic polymers: A class of novel material. *J. Polym. Mater*.14:105.

Karger, J., and D.M. Ruthven. 1992. *Diffusion in zeolites*. New York: J. Wiley. Krätschmer, W., L.D. Lamb, K. Fostiropoulos, and D.R. Huffman. 1990. *Nature* 347:354.

Karplus, M. & Weaver, D. L. (1976) *Nature (London)* 260, 404-406.

Kasuga, T., M. Hiramatsu, A. Hoson, T. Sekino, and K. Niihara. 1998. *Langmuir* 14:3160.

Katari, J.E.B., V.L. Colvin, and A.P. Alivisatos. 1994. *J. Phys. Chem*. 98:4109.

Ke, M., S.A. Hackney, W.W. Milligan, and E.C. Aifantis. 1995. *Nanostructured Mater*. 5:689.

Kear, B., and G. Skandan. 1998. Nanostructural bulk materials: Synthesis, processing, properties and performance. *In R&D status and trends*, ed.

Kear, B.H., R.K. Sadangi, and S.C. Liao. 1997. Synthesis of WC/Co/diamond nanocomposites. In *Proc. of the Joint NSF-NIST Conf. on Nanoparticles*.

Kirschvink, J.L., A. Koyayashi-Kirschvink, and B.J. Woodford. 1992. Magnetite biomineralization in the human brain. *Proc. Nat'l. Acad. Sci.* USA 89:7683-7687.

Kishida, M., T. Fujita, K. Umakoshi, J. Ishiyama, H. Nagata, and K. Wakabayashi. 1995. *Chem. Commun*. 763.

Klein, J.D., et al. 1993. *Chem. Mater*. 5:902. Koch, C.C. 1989. Materials synthesis by mechanical alloying. *Annual Review of Mater. Sci.* 19:121-143.

Knight, F.H. Risk, Uncertainty and Profit. London School of Economics and Political Science, London, 1933.

Koch, C.C. 1998. Bulk behavior. In *R&D status and trends*, ed. Siegel et al. 32 *Evelyn L. Hu and David T. Shaw.*

Kortan, A.R., R. Hull, R.L. Opila, M.G. Bawendi, M.L. Steigerwald, P.J. Carroll, and L.E. Brus. 1990. *J. Am. Chem. Soc.* 112:1327.

Kourilsky, Ph.&Viney, G. Le Principe de précaution. Report to the Prime Minister, Paris, Éditions Odile Jacob, 2000.

Krejchi, M.T., E.D.T. Atkins, A.J. Waddon, M.J. Fournier, T.L. Mason, and D.A. Tirrell. 1994. Chemical sequence control of beta-sheet assembly in macromolecular crystals of periodic proteins. *Science* 265:1427-1432.

Krejchi, M.T., S.J. Cooper, Y. Deguchi, E.D.T. Atkins, M.J. Fournier, T.L. Mason, and D.A. Tirrell. 1997. Crystal structures of chain-folded antiparallel beta-sheet assemblies from sequence-designed periodic polypeptides. *Macromolecules* 30:5012-5024.

Kresge, C.T., M.E. Leonowicz, W.J. Roth, J.C. Vartuli, and J.S. Beck. 1992. *Nature* 359:710. Kupperman, A., S. Nadimi, S. Oliver, G. Ozin, J. Garcés, and M. Olken. 1993. *Nature* 365:239.

Kroes, P. & Meijers, A. "The Dual Nature of Technical Artifacts—Presentation of a New Research Programme." Techné 6, 2 (2002): 4-8.

Kuhn, T.: 1970, *The Structure of Scientific Revolutions* (second edition), University of Chicago Press, Chicago.

Kumar, A., and G.M. Whitesides. 1993. Features of gold having micrometer to centimeter dimensions can be formed through a combination of staming with an elastomeric stamp and an alkanethiol ink followed by chemical etching. *App. Phys. Lett.* 63:2002-2004.

Kurzweil, R.: 1998, *The Age of Spiritual Machines, How We Will Live, Work and Think in the New Age of Intelligent Machines*, Phoenix, New York.

Kyprianidou-Leodidou, T., W. Caseri, and V. Suter. 1994. *J. Phys. Chem.* 98:8992. Kung, H.H., and E.I. Ko, 1996. *Chem. Eng. J.* 64:203.

L. J. Perry and R. Wetzel, "Disulfide bond engineered into T4 lysozyme: stabilization of the protein toward thermal inactivation." *Science,* 226: 555-557 (1984).

L. Regan and W. F. DeGrado, "Characterization of a helical protein designed from first principles." *Science,* 241: 976-978 (1988).

Laudan, L.: 1984, *Science and Values: The Aims of Science and their Role in Scientific Debate*, University of California Press, Berkeley.

Lehn, J.M., 2004: 'Une chimie supramoléculaire foisonnante', *La lettre de l'Académie des sciences*, 10, 12-13.

Leonard, D., M. Krishnamurthy, C.M. Reaves, S.P. Denbaars, and P.M. Petroff. 1993. Direct formation of quantum-sized dots from uniform coherent islands of InGaAs on GaAs surfaces. *Appl. Phys. Lett.* 63:3203-5.

Liang, G., Z. Li, and E. Wang. 1996. *J. Mater. Sci.* Makino, A., A. Inoue, T. Hatanai, and T. Bitoh. 1997. *Materials Science Forum* 235-238: 723.

Lide, D.R., ed. 1993-1994. *CRC Handbook of Chemistry and Physics*, 74th ed. Lin, H.-P., and C.-Y. Mou. 1996. *Science* 273:765.

Lukens, R.: 2004, 'Historic Nanotechnology', *Chemical Heritage*, 22, 29.

Majetich, S.A. and A.C. Canter. 1993. *J. Phys. Chem.* 97:8727.

Mao, C., W. Sun, and N.C. Seeman. 1997. Construction of Borromean rings from DNA. *Nature* 386:137-138.

Martin, T.P., N. Malinowski, U. Zimmerman, U. Naher, and H. Schaber. 1993. *J. Chem. Phys.* 99:4210.

Martin, T.P., U. Naher, H. Schaber, U. Zimmerman. 1993. *Phys. Rev. Lett.* 70:3079. Prigogine, I., and S. Rice. 1988. *Advances in chemical physics*, Vol. 70, Parts 1 & 2. New York: J. Wiley.

McCammon, J. A., Gelin, B. R. & Karplus, M. (1977) *Nature (London)* 267, 585-590.

McConnell, H.M. 1996. Light-addressable potentiometric sensor: Applications to drug discovery. In *Nanofabrication and biosystems*, ed. Hoch et al.

McCulloch, W.S.: 1962, 'The imitation of one form of life by another – Biomimesis', in: E.E. Bernard & M.R. Kare (eds.), *Biological prototypes and synthetic systems*, Plenum Press, New York, vol. 1., p. 393-97.

Mehl, R.F., and R.W. Cahn. 1983. Historical development. In *Physical metallurgy*. North Holland. Milligan, W.W., S.A. Hackney, M. Ke, and E.C. Aifantis. 1993. *Nanostructured Materials* 2:267.

Minsky, M.: 1995, 'Virtual Molecular Reality', in: Krummenacker, M. & Lewis, J. (eds.), *Prospects in Nanotechology. Proceedings of the 1st general conference on nanotechnology: developments, applications, and opportunities, November 11-14, 1992, Palo-Alto*, John Wiley & Sons, New York, pp. 187-205.

Mishra, R.S., and A.K. Mukherjee. 1997. Oral presentation at TMS meeting, Indianapolis, Indiana, 16-18 September 1997, to be published in proceedings of Symp. "Mechanical Behavior of Bulk Nano-Materials."

Mishra, R.S., R.Z. Valiev, and A.K. Mukherjee. 1997. *Nanostructured Materials* 9:473. Morris, D.G., and M.A. Morris. 1991. *Acta Metall. Mater*. 39:1763-1779.

Moore, J.C., H.M. Jin, O. Kuchner, and F.H. Arnold. 1997. Strategies for the in vitro evolution of protein function: Enzyme evolution by random recombination of improved sequences. *J. Mol. Biol*. 272:336-347.

Morris, D.G., and M.A. Morris. 1997. *Materials Science Forum* 235-238:861. Nagpal, P., and I. Baker. 1990. *Scripta Metall. Mater*. 24:2381.

Nieman, G.W., J.R. Weertman, and R.W. Siegel. 1991a. *Mater. Res. Soc. Symp. Proc*. 206:581-586.

Nomura, M. & Held, W. (1974) in *Ribosomes*, eds. Nomura, M., Tissiers, A. & Lengyel, P. (Cold Spring Harbor Laboratory, Cold Spring Harbor, NY), pp. 193-203.

Nordmann, A. (rapp.) Converging Technologies—Shaping the Future of European Societies, European Commission report, 2004.

Parker, J.C., et al. 1995. U.S. Patent 5,460,701. POST (Parliamentary Office of Science and Technology). 1995.

Phoenix, C.; Drexler, E.: 2004, 'Safe Exponential Manufacturing', *Nanotechnology*, 15, 869-72.

Rademann, K., B. Kaiser, U. Even, F. Hensel. 1987. *Phys. Rev. Lett*. 59:2319. Rao, C.N.R., B.C. Satishkumar, and A. Govindaraj. 1997. *Chem. Commun*. 1581.

Ramanan, V.R. 1998. Nanocrystalline soft magnetic alloys for applications in electrical and electronic devices. In *R&D Status and Trends*, ed. Siegel et al.

Rao, M.B., and S. Sircar. 1993. *Gas Separation and Purification* 7:279. Reetz, M.T., et al. 1995. *Science* 267:367.

Rechard, R.P. "Historical Relationship Between Performance Assessment for Radioactive Waste Disposal and Other Types of Risk Assessment." Risk Analysis 19, 5 (1999): 763-807.

Rietman, E.A.: 2001, 'Drexler hypothesis of a universal assembler is supported not by theoretical arguments alone but by existence proof in the form of biological life', in: *Molecular Engineering of Nanosystems*, Springer, New York & Berlin.

Rip, A., Misa, Th. J.,&Schot, J. W. (eds.). Managing Technology in Society. The Approach of Constructive Technology Assessment. London: Pinter Publishers, 1995.

Roco, M.&Bainbridge, W. (eds.). Converging Technologies for Improving Human Performances, National Science Foundation report, 2002.

Rofagha, R., R. Langer, A.M. El-Sherik, U. Erb, G. Palumbo, and K.T. Aust. 1991. *Scr. Metall. Mater*. 25:2867.

Roher, H. 1993. *Jpn. J. Appl. Phys*. 32:1335. Rohlfing, E.A., D.M. Cox, and A. Kaldor. 1984. *J. Chem. Phys*. 81:3846.

Romanov, A.E., V.I. Vladimirov. 1992. In *Dislocations in solids*, ed. F.R.N. Nabarro, Vol. 9. Amsterdam: North-Holland. Salishekev, G.A., O.R. Valiakhmetov, V.A. Valitov, and S.K. Mukhtarov. 1994. *Materials Science Forum*. 170-172:121.

Rosenberg, N. "Why Technology Forecasts Often Fail." Futurist July/August 1995, 16- 21.

Ruthven, D.M., S. Farooq, K.S. Knaebel. 1994. *Pressure swing adsorption*. New York: VCH Publishers. Sayari, A. 1996. *Chem. Mater*. 8:1840.

S. P. Ho and W. F. DeGrado, "Design of a 4-Helix bundle protein: Synthesis of peptides which self-associate into a helical protein." *J. Am. Chem. Soc.,* 109: 6751-6758 (1987).

Sanders, P.G., J.A. Eastman, and J.R. Weertman. 1996. In *Processing and properties of nanocrystalline materials*, ed. Suryanarayana et al.

Sanders, P.G., M. Rittner, E. Kiedaisch, J.R. Weertman, H. Kung, and Y.C. Lu. 1997. *Nanostructured Mater*. 9:433.

Sankey, H.: 1994, *The Incommensurability Thesis*, Avebury, Aldershot.

Sarikaya, M.; Aksay, I. (eds.), 1995, *Biomimetics. Design and Processing of Materials*, AIP Press, Woodbury, New York.

Scanlan, R.M., W.A. Fietz, and E.F. Koch. 1975. *J. Appl. Phys*. 46:2244.

Scheraga, H. A. (1978) in *Versatilty of Proteins*, ed. Li, C. H. (Academic, New York), pp. 119-132.

Schiefsky, M.J.: forthcoming, 'Art and Nature in Ancient Mechanics', in: W.R. Newman & B. Bensaude-Vincent (eds.), *The Artificial and the Natural: An Ancient Debate and its Modern Descendants*, MIT Press, Cambridge, MA.

Schultz, L., J. Wecker, and E. Hellstern. 1987. *J. Appl. Phys*. 61:3583.

Schultz, L., K. Schnitzke, and J. Wecker. 1989. *J. Magn. Mater*. 80:115.

Schwarz, R.B. 1998. Storage of hydrogen powders with nanosized crystalline domains. In *R&D Status and Trends*, ed. Siegel et al.

Shen, T.D., C.C. Koch, T.Y. Tsui, and G.M. Pharr. 1995. *J. Mater. Res.* 10:2892.

Shull, R.D. 1998. NIST activities in nanotechnology. In *R&D Status and Trends*, ed.

Shull, R.D., R.D. McMichael, and J.J. Ritter. 1993. *Nanostructured Mater*. 2:205.

Siegel et al. Ogunnaike, B and W. Ray. 1994. *Process dynamics, modeling and control*. Oxford University Press, pp 5-21; 1033-48.

Tighe, T.S., J.M. Worlock, and M.L. Roukes. 1997. Direct thermal conductance measurements on suspended monocrystalline nanostructures. *Appl. Phys. Lett*. 70:2687-9.

3

Tools and Techniques of Nano Research

Tools for Nano-technology: An Introduction

Electron Microscopy

In 1924 L. de BROGLIE discovered the wave-character of electron rays thus giving the prerequisite for the construction of the electron microscope. The prototype was built by M. KNOLL and E. RUSKA (Technische Universität Berlin, 1932). One of the first biological objects depicted was the tobacco mosaic virus (TMV). The first picture of a cell was published in 1945 by K. R. PORTER, A. CLAUDE and E. F. FULLAM (Rockefeller Institute, New York).

The Transmission Electron Microscope (TEM)

The conventional electron microscopy is nowadays called TEM (transmission electron microscopy). We will therefore start with its construction. The ray of electrons is produced by a pin-shaped cathode heated up by current. The electrons are vacuumed up by a high voltage at the anode. The acceleration voltage is between 50 and 150 kV. The higher it is, the shorter are the electron waves and the higher is the power of resolution. But this factor is hardly ever limiting. The power of resolution of electron microscopy is usually restrained by the quality of the lens-systems and especially by the technique with which the preparation has been achieved. Modern gadgets have powers of resolution that range from 0,2 - 0,3 nm. The useful resolution is therefore around 300,000 x.

The accelerated ray of electrons passes a drill-hole at the bottom of the anode. Its following way is analogous to that of a ray of light in a light microscope. The lens-systems consist of electronic coils generating an electromagnetic field. The ray is first focused by a condenser. It then passes through the object, where it is partially deflected. The degree of deflection depends on the electron density of the object. The greater the mass of the atoms, the greater is the degree of deflection. Biological objects have only weak contrasts since they consist mainly of atoms with low atomic numbers (C, H, N, O). Consequently it is necessary to treat the preparations with special contrast enhancing chemicals (heavy metals) to get at least some contrast. Additionally they are not to be thicker than 100 nm, because the temperature is raising due to electron absorption. This again can lead to destruction of the preparation. It is generally impossible to examine living objects.

After passing the object the scattered electrons are collected by an objective. Thereby an image is formed, that is subsequently enlarged by an additional lens-system (called projective with electron microscopes). The thus formed image is made visible on a fluorescent screen or it is documented on photographic material. Photos taken with electron microscopes are always black and white. The degree of darkness corresponds to the electron density (= differences in atom masses) of the candled preparation.

The Scanning Rlectron Microscope (SEM)

The path of the electron beam within the scanning electron microscope differs from that of the TEM. The technology used is based on television techniques. The method is suitable for the depiction of preparations with conductive surfaces. Biological objects have thus to be made conductive by coating with a thin layer of heavy metal (usually gold is taken). The power of resolution is normally smaller than in transmission electron microscopes, but the depth of focus is several orders of magnitude greater. Scanning electron microscopy is therefore also well-suited for very low magnifications. Numerous examples will be given in the following.

The surface of the object is scanned with the electron beam point by point whereby secondary electrons are set free. The intensity of this secondary radiation is dependent on the angle of inclination of the object's surface. The secondary electrons are collected by a detector that sits at an angle at the side above the object. The signal is then enhanced electronically. The magnification can be chosen smoothly (depending on the model) and the image appears a little later on a viewing screen. The properties of the light microscope as opposed to that of transmission and scanning electron microscopes are collected in a table. Finally some outlines of new and further developments are given. The high voltage electron microscope: it operates with an accelerating voltage of 700 - 3000 kV. Its power of resolution is greater, the preparation can be thicker, the strain on the preparation is smaller. But the enormous technical expenditure is disadvantageous. Only few gadgets exist. New results concerning botany have not been gained. The scanning transmission electron microscope (STEM): In this development of the SEM do the electrons pass through the preparation and the secondary radiation thus generated is used for image formation. Here, too, the expenditure is large, but it is still worthwhile, since large molecules like nucleic acids or proteins or molecular complexes like viruses can be depicted much better and gentler than with the TEM. No news for botany, though. The interpretation of images gained with electron microscopy is increasingly done with computerized interpretation programs. But they are usually only suitable for the reconstruction of regularly recurring patterns and these, again, are found more often on a molecular level than on a cellular one.

Atomic Force Microscopy—AFM tools

The Company: Since the late 1990's Novascan has been a nanotechnology company that specializes in atomic force microscope products (AFM) and AFM / SPM services. Our tools are used worldwide to visualize, characterize and manipulate microscopic and nanoscopic environments.

Our Mission: We are committed to developing and producing innnovative and state of the art devices that facilitate our clients projects in nanotechnology. These improvements range from specialized AFM instruments to the probes and sensors used to characterize samples.

Technology: At Novascan we strive to develop forward looking devices with the hope that we can help you make new scientific discoveries in your work and projects. Want to learn more?

Beneficial Nanotechnology

Foresight is the leading think tank and public interest institute on nanotechnology. Founded in 1986, Foresight was the first organization to educate society about the benefits and risks of nanotechnology. At that time, nanotechnology was a little-known concept. Today, with the basic framework of public understanding in place, we are refocusing our efforts on guiding nanotechnology research, public policy and education to address the critical challenges facing humanity. Foresight's new mission is to ensure the beneficial implementation of nanotechnology. Foresight is accomplishing this by providing balanced, accurate and timely information to help society understand and utilize nanotechnology through public policy activities, publications, guidelines, networking events, tutorials, conferences, roadmaps and prizes. Foresight is a member-supported organization. Our membership, including over 14,000 individuals and a growing number of corporations, is diverse demographically and geographically. They are interested in ensuring that the future of nanotechnology unfolds for the benefit of all. These concerned individuals include scientists, engineers, business people, investors, publishers, artists, ethicists, policy makers, interested laypersons, and students from grammar school to graduate level.

Menlo Park, CA—August 29, 2005—Floyd Kvamme, Co-Chair of the President's Council of Advisors on Science and Technology (PCAST), and Partner at Kleiner Perkins Caufield & Byers, will present his view on Washington policies and nanotechnology at the 13th Foresight Conference: Advancing Beneficial Nanotechnology: Focusing on the Cutting Edge, to be held October 22-27, 2005 at the San Francisco Airport Marriott.

Menlo Park, CA—July 27, 2005—Foresight Nanotech Institute, the original organization in the nanotechnology field, has appointed commercial space flight pioneer, Dr. Peter Diamandis, CEO of the X PRIZE Foundation and CEO of Zero Gravity Corporation; venture capitalist and media expert, Ed Niehaus, Partner at Cypress Ventures; and nanotechnology leader and executive, James Von Ehr II, Founder and CEO of Zyvex Corporation, to its Board of Directors.

While nanotechnology has made great advances in the last two decades, it has yet to fulfill its ultimate potential. Foresight is dedicated to fostering nanotechnologies that can make a significant contribution to solving critical challenges which humanity faces— *Foresight Nanotechnology Challenges.* These challenges are:

1. Meeting global energy needs with clean solutions
2. Providing abundant clean water globally
3. Increasing the health and longevity of human life
4. Maximizing the productivity of agriculture
5. Making powerful information technology available everywhere
6. Enabling the development of space

Nanotechnology Challenges

Balancing humanity's energy demands while protecting the environment is a major challenge. Nanotechnology will help to solve the dilemma of energy needs and limited planetary resources through more efficient generation, storage and distribution.

The demand for fresh water is increasing. Considering the current rate of consumption and projected population growth, some two-thirds of the world will be affected by drought by the year 2050. Nanotechnology can help solve this problem through improved water purification and filtration.

Humans are living longer lives, yet infectious diseases and cancer continue to kill millions annually. Because of an aging population there could be a 50% increase of new cancer cases by the year 2020. Nanotechnology will enhance the quality of life for human beings through medical diagnostics, drug delivery and customized therapy.

Pressure on the world's food sources is ever increasing while harvests have fallen short in recent years. It is anticipated that our world population will swell to 8.9 billion by the year 2050 putting even greater demands on agriculture. Precision farming, targeted pest management and the creation of high yield crops are a few nanotech solutions.

Humanity will need to cooperate as we respond to disasters and critical threats to our survival. A "planetary nervous system" fostering rapid communication and cross-cultural relationships is needed. Nanotechnology applications in electronics will increase access through reduced cost and higher performance of memory, networks, processors and components.

Heavy demands on resources and raw materials are creating challenges on earth, whereas these items are plentiful in space. Current obstacles to developing space are cost, reliability, safety, and performance. Nanotechnology will solve these through improved fuels, smart materials, uniforms and environments.

Nanotechnology: Societal Interactions and Implications

New technologies come into being through a complex interplay of technical and social factors. The process of innovation that will produce nanotechnology and diffuse its benefits into society is complex and only partially understood. Economists, as well as scholars in other fields, have long studied the generation, diffusion, and impact of scientific and technological innovation. These studies outline the variables likely to determine the rate and direction of these impacts, and to identify relevant research questions. They provide a foundation on which to build studies of societal implications of nanotechnology. Scientific discoveries do not generally change society directly; they can set the stage for the change that comes about through the confluence of old and new technologies in a context of evolving economic and social needs. The thorough diffusion of even major new developments rarely happens all at once. Nanotechnologies are so diverse that their manifold effects will likely take decades to work their way through the socio-economic system. While market factors will determine ultimately the rate at which advances in nanotechnology get commercialized, sustained support for nanoscience research is necessary in this early stage of development so as not to become a rate-limiting factor. Expediting research (innovation) and its incorporation into beneficial technology is a major challenge to the NNI.

Perhaps the greatest difficulty in predicting the societal impacts of new technologies has to do with the fact that once the technical and commercial feasibility of an innovation is demonstrated, subsequent developments may be as much in the hands of users as in those of the innovators. The diffusion and impact of technological innovations often depends on the development of complementary technologies and of the user network. As a result, new technologies can affect society in ways that were not intended by those who initiated Societal Implications of Nanoscience and Nanotechnology them. Often these unintended consequences are beneficial, such as spin-offs with valuable applications in fields remote from the original innovation. For instance, consider how the Internet has progressed from a technology supported by the Department of Defense's Advanced Research Projects Agency (DARPA) to facilitate digital communications among universities with DARPA contracts, to a means by which teenagers and college students exchange music files. In another example, intended benefits may also have unintended or "second-order consequences." Nanotechnologybased medical treatments, for example, may significantly improve life span and quality of life for elderly people; a second-order consequence would be an increase in the proportion of the population that is elderly, which might require changes in pensions or health insurance, an increase in the retirement age, or a substantial increase in the secondary careers undertaken by older people. Another potential consequence that would need to be addressed is the potential increase of inequality in the distribution of wealth that we may call the "nano divide." Those who participate in the "nano revolution" stand to become very wealthy. Those who do not may find increasingly difficult to afford the technological wonders that it engenders. One near-term example will be in medical care: nanotech-based treatments may be initially expensive, hence accessible only to the very rich.

Other consequences are not so desirable, such as the risk of closing old industries and environmental pollution, which sometimes becomes a problem, especially for largescale technologies. To assess a nanotechnology (or any technology) in terms of its unintended consequences, researchers must examine the entire system of which the technology is a part through its entire life cycle. As the case of electric automobiles illustrates, without a careful analysis of the entire set of activities that produce, operate, and eventually dispose of a technology, people may leap to false conclusions about the extent to which the technology pollutes. For example, manufacture and disposal of an electric vehicle's battery may release more lead into the environment than if the vehicle had been fueled throughout its working life by leaded gasoline. One concern about nanotechnology's unintended consequences raises the question of the uncontrolled development of self-replicating nanoscale machines. A number of very serious technical challenges would have to be overcome before it would be possible to create nanoscale machines that could reproduce themselves in the natural environment. Some of these challenges appear to be insurmountable with respect to chemistry and physical principles, and it may be technically impossible to create self-reproducing mechanical nanoscale robots of the sort that some visionaries have imagined. A new form of life different from that known (i.e., carbon-based) would be a dramatic change that is not foreseen in the near future.

Initially, the impact of nanotechnology will likely be limited to a few specific products and services. Nanotechnology-based goods and services will probably be introduced earlier to those markets where consumers are willing to pay a premium for new or improved performance. Such primary effects would be to make things work better, cheaper, with more features, etc. This might, for example, increase food yields, generate new textiles for clothing, improve power production, or cure a certain disease. As mentioned above, by and large, the displacement of an old technology by a new one tends Societal Implications of Nanoscience and Nanotechnology to be both slow and incomplete. As a result, nanotechnology will coexist for a long time with older technologies rather than suddenly displacing them. During that time it will affect the further development of those competing technologies. Other secondary effects might be shifts in demand for products and services, so that people come to expect different kinds of food, medical care, entertainment, etc. This shift in demand may also initiate a tertiary effect, the need for augmented nanotechnology infrastructure — interdisciplinary research centers, new educational programs to supply nanoscientists and nanotechnologists, etc. Other tertiary effects would move upstream in our social structures and cultural patterns, such as shifts in education and career patterns, family life, government structure, and so forth. *While there is no way of knowing, a priori, the unintended and higher order consequences of nanotechnology, the participation of social scientists in the NNI may allow for important issues to be identified earlier, the right questions to be raised, and necessary corrective actions taken.* An effective and cost-efficient way to protect the public and deal with nanotechnology's potential negative consequences is to develop a tradition of social-science-based countermeasures — and to support research in publicly recognized institutions on the processes that develop nanotechnology and apply it in diverse areas of life.

An important aim of a societal impact investigation of nanotechnology is to identify harms, conflicts over justice and fairness, and issues concerning respect for persons. For example, changes in workforce needs and human resources are likely to bring benefits to some and harm to others. Other examples of potential issues include safeguards for workers engaged with hazardous production processes, equity disputes raised by intellectual property protection, and questions about relationships between government, industry, and universities. Scientists and engineers bring to their work a laudable concern for the social value of their labors. However, those working in a particular technical field may be focused on the immediate technical challenges and not see all of the potential social and ethical implications. It is important to include a wide range of interests, values, and perspectives in the overall decision process that charts the future development of nanotechnology. Involvement of members of the public or their representatives has the added benefit of respecting their interests and enlisting their support.

The inclusion of social scientists and humanistic scholars, such as philosophers of ethics, in the social process of setting visions for nanotechnology is an important step for the NNI. As scientists or dedicated scholars in their own right, they can respect the professional integrity of nanoscientists and nanotechnologists, while contributing a fresh perspective. Given appropriate support, they could inform themselves deeply enough about a particular nanotechnology to have a well-grounded evaluation. At the same time, they are professionally trained representatives of the public interest and capable of functioning as communicators between nanotechnologists and the public or government officials. Their input may help maximize the societal benefits of the technology while reducing the possibility of debilitating public controversies. In addition, attention needs to be given to the individual responsibility of engineers, scientists, and others involved in the processes of generating powerful new nanotechnologies. Professional societies have a role to play in providing opportunities for discussing and devising guidelines that incorporate relevant ethical principles into emerging issues. Perhaps most importantly, ethics must be incorporated effectively into the curriculum for training new nanoscientists, nanotechnologists, and nanofabrication technicians.

The United States faces the daunting challenge of attracting enough of the best graduate students to the physical sciences and engineering disciplines. Under present conditions, far too few good students are attracted to the fields relevant to nanotechnology. To some extent, this is a problem faced by all of the sciences, but the problem is especially acute for nanotechnology because a very large number of talented scientists, engineers, and technicians will be needed to build the nanotechnology industries of the future, and these professionals will require an interdisciplinary perspective. Development of nanotechnology will depend upon multidisciplinary teams of highly trained people with backgrounds in biology, medicine, applied and computational mathematics, physics, chemistry, and in electrical, chemical, and mechanical engineering. Team leaders and innovators will probably need expertise in multiple subsets of these disciplines, and all members of the team will need a general appreciation of the other members' fields. Developing a broadly trained and educated workforce presents a severe challenge to our four-year degree and two-year degree educational institutions, which favor compartmentalized learning. Because current educational trends favor specialization, there must be fundamental changes in our educational systems. However, introducing new degree programs in nanotechnology that provide a shallow overview of many disciplines, none in sufficient depth to make major contributions, may not give students the training that is needed to meet the future challenges. The right balance between specialization and interdisciplinary training needs to be worked out through innovative demonstration programs and research on the education process and workforce needs. Education in nanoscience and nanotechnology requires special laboratory facilities that can be quite expensive. Given the cost of creating and sustaining such facilities, their incorporation into nanotechnology workforce development presents a considerable challenge. Under the present education system, many engineering schools, let alone the two-year-degree colleges, cannot offer students any exposure to the practice of nanofabrication. Innovative solutions will have to be found, such as new partnerships with industry and the establishment of nanofabrication facilities that are shared by consortia of colleges, universities, and engineering schools. Web-based, remote access to those facilities may provide a powerful new approach not available previously. Despite the tremendous educational challenges, the exciting intellectual, economic, and social opportunities of nanotechnology might become a major factor in reinvigorating our nation's youth for careers in science and technology.

A related educational challenge is the very small number of social scientists who have the technical background and research orientation that would allow them to conduct competent research on the societal implications of nanotechnology. At the university level, liberal arts education gives far too low a priority to scientific literacy. Social science professional societies, universities, and government agencies will have to make a long-term commitment to attract talented young social scientists to this area of research and to encourage them to gain the necessary professional skills and awareness of nanotechnology. This will require research on the societal implications of nanotechnology at a consistent and high enough level to establish this as a viable field of social science research.

Tools & Techinques of Nanoparticles, Nano-material & Nano-structuring

One third of the tubes are metallic or quasi, the others are semiconductors with moderate band-gaps. If solid particles become small enough, in the nanometer range, they become electronically comparable to atoms and molecules. Instead of the laws of classical physics, they become quantum systems with discrete energy states. This is the basis for a number of useful applications. Quantum dots are reminiscent of the rhetorical question raised by Greek philosophers in antiquity: consider a sand-heap. If one removes from it a single grain of sand, obviously it is still a sand-heap. One continues doing so. At which stage does the sand-heap become something else? One of the most exciting applications of quantum dots is the single electron tunneling transistor. An electrometer measures electric charge. A single electron transistor is, at present, the most sensitive electrometer. Small alterations of the charge distribution beneath a small isolated island of aluminum atoms markedly affect the conductivity across that island. When a gate voltage is applied to an electron donor material, such as tellurium, placed underneath the island, changes in conductivity can be monitored. Nanocrystals are indeed important in single electron devices operating at room temperature, such as supersensitive electrometers and memory storage. Capped nanocrystals of both metals and semiconductors by virtue of their size possess capacitance in the range of attofarads (1 aF = 10^{-18} F). Charging a nanocrystal with an extra electron perturbs it to such an extent that the next electron requires an appreciable change in the charging potential. This is often seen as a 'Coulomb staircase' in the current-voltage tunneling spectra. Indeed, the charging energy varies linearly with the inverse of the diameter of the nanocrystal. Such sensitivity to single electron charging makes a nanocrystal an ideal candidate for use in single electron transistors and memory devices. There are metallic and there are semiconducting nanoparticles. The former enjoy industrial and medical applications, for their magnetic properties in particular. The latter have uses in electronics and optoelectronics. One can also write with nanoparticles, dip-pen lithography is one of the techniques. More generally, the challenge is to organize nanoparticles in three dimensions. The natural and the artificial once again meet! Biosystems, such as enzymes, antibodies, and nucleic acids have sizes commensurate with those of man-made nanoparticles. Hence, it becomes tempting to marry and hybridize them into nanocomposites, which is one of the most recent developments. Two of the main thrusts of nanotechnology have been self-assembled ordered arrays of atoms with specific properties (electrical conductivity, magnetism, *etc.*) and well-defined, molecule-like clusters of atoms endowed with specific properties. The keyword here is specificity.

The standard account of the birth of nanotechnologies dates it to the lecture by Richard P. Feynman on December 29, 1959 at the annual meeting of the American Physical Society held at Caltech. The books under review rightfully salute Feynman as the Founding Father of the field. As is well-known, the brilliant Caltech physicist ended his talk with a dare, announcing two prizes financed with his own money for the miniaturization of the pages of a book and for the building of a miniscule electric motor, only 1/64 inch cube. Feynman's lecture makes interesting reading. Its perspective is identical to that in Jules Verne's books, not visionary in the sense of a prophecy, imagining revolutionary developments and devices. The outlook is much more down to earth. Jules Verne and Richard P. Feynman took heed of recent scientific and technological advances to extrapolate to what the future would almost certainly witness, in a prospective. In his lecture Feynman took as his starting point the feasibility of scaling down the contents of the world's libraries. He guesstimated their holdings at about 24 million volumes. He figured that their total information could be contained on an area of about 35 pages of the *Encyclopaedia Britannica*. The discussions of potential economic, medical, and environmental benefits may have given the false impression that nanotechnology will create a wondrous utopia in which all human problems are solved and we all live happily ever after. This is even more mistaken than the idea that new technologies always cause more problems than they solve. Many of the main constraints and difficulties faced by people are based not on technology or its lack, but instead by the very nature of the world we live in and the essence of our humanness. Increasing affluence based on molecular manufacturing won't end economic problems any more than past increases in affluence have. Wilderness can still be destroyed; people can be oppressed; financial markets can be unstable; trade wars can be waged; inflation can soar; individuals, companies, and nations can go into debt; bureaucracy can stifle innovation; tax levels can become crippling; wars and terrorism can rage. None of these will automatically be stopped by advanced technology. What is more, the potential benefits of new technologies aren't automatic. Nanotechnology could be used to restore the environment, to spread wealth, and to cure most illness. But will it? This depends on human action, working within the limits set by the real world. This chapter first describes some of the limits to what nanotechnology can accomplish, and then some of the adverse side effects of its basically good applications. The next will discuss the problem of accidents, which seems manageable, and then the far greater problem of potential abuse of new capabilities.

The world imposes limits on what we can do. Technology in general (and nanotechnology in particular) can provide padding for us as we throw ourselves against these hard, sharp limitations, and can sometimes help us slip past old limits through previously unknown gaps. Eventually, though, we will encounter new limits. In the end, solid constraints will limit human action no matter how much we juggle atoms and molecules, or the bits and bytes of information. Let's look at some of these, starting with the most abstract and long term—the most definite and hardest to avoid—and moving toward the more personal and near term. Many problems differ fundamentally from the material problems of limited matter and energy: they involve information. Some of the most precious stores of information in the world today are the genetic codes of the biosphere. This information, different for virtually every individual organism, is the product of millions of events that we are incapable of modeling or recreating. When this information is lost, it is lost forever. When the atoms encoding this information are thoroughly scattered, there seems to be no way to retrieve it. With any species, most genetic information is shared in common, found in all members of that species. But the variations in genetic code between individuals are important, both to the individuals themselves and to the health and prospects of the species as a whole. Consider the northern white rhino, whose numbers have dropped to an estimated thirty-two animals, or the California condor, of which only forty remain, all in captivity. Even if biologists succeed in reestablishing these species—eight condors were hatched in 1989—much of the diversity of their genetic information has been lost. Worse yet are extinctions of species for which no tissue samples were saved. The future may see some amazing recoveries: dry skin and bones may yield a complete set of genes when sifted by molecular machinery, and even current techniques have been used to recover genes from an ancient leaf, almost 20 million years old. Our eyes and instruments cannot yet tell us how much information from the past remains, but we do know that genetic information is being lost every day, and once lost, it is irretrievable.

People have often been wrong about physical limits, confusing the limits of their technology with the limits of the possible. As a result, learned men first dismissed the idea of heavier-than-air flight, and then dismissed the idea of flying to the Moon. Yet physical limits are real, and all technology—past, present, and future—will stay within those limits. There is even reason to suspect that some of those limits are where the learned now believe them to be. Nanotechnology will make it possible to push closer to the real limits set by natural law, but it will not change those laws or the limits they set. It will not affect the law of gravity, the gravitational constant, the speed of light, the charge of the electron, the radius of the hydrogen atom, the value of Planck's constant, the effects of the uncertainty principle, the principle of least action, the mass of the proton, the laws of thermodynamics, or the boiling point of water. Nanotechnology won't make energy or matter from nothing. It seems a good bet that no one will build a faster than light spacecraft, or an antigravity machine, or a cable twice as strong as diamond. There are limits. Science today may be wrong about some limits, but scientific knowledge is practically defined to be our best information about how the world works, so it isn't wise to bet against it. There will be claims that nanotechnology will be able to do things that it can't, or that capabilities are around the corner when they aren't. Sometimes these will be innocent errors, sometimes they will be culpably stupid errors, and sometimes they will be what amounts to fraud. Among the problems that nanotechnology cannot solve is that of misguided claims, by people calling themselves "scientists," "engineers," or "businesspeople," that they have a big technical breakthrough worth a fortune. Every interesting new technology, particularly in its early days, is a chaotic mix of competent workers and charlatans.

For every Thomas Edison inventing useful products such as light bulbs or the precursor of movie projectors, there were people promoting electric hairbrushes to cure baldness, and electric shoes, electric belts, electric hats—the list goes on—that authoritatively claimed cures for infertility, overweight, underweight, and all the ills and discomforts of mankind. Today, we laugh at the credulity of our forefathers who bought these gadgets; we shouldn't, unless we laugh at our own times as well.

Natural law imposes limits, but so does the nature of human beings. These will continue as long as people do. Reproduction is a deeply ingrained instinct enforced by the march of time, which ruthlessly discards the genetic material of all who neglect it. Many would argue that the Earth is already overpopulated. While nanotechnology could enable the current population, and even a greatly increased one, to live more lightly on the Earth, there will still be limits to Earth's capacity. The norms of human life are shaped by ancient patterns: high rates of infant or childhood mortality have been facts of life for millennia, and having many, many children has been a way to ensure that one or two will survive to work on the farm, and to care for you in your old age. Large families naturally become traditional. When modern medicine and reliable food supplies change those conditions—as they have, in cultural terms, virtually overnight—behavior does not shift as quickly. The result is the Third World population boom. In Western countries, where there has been time for behavior to adapt, a huge family is the exception. It might seem that our problem is solved. Molecular manufacturing can make everyone wealthy, and wealthy populations today have stable or shrinking populations. The Earth can support more people with advanced technologies, and these will also open up the vast room and resources of the world beyond Earth. Would that this were true. If 99 per cent of the people in a population respond to wealth by reducing childbearing, the population will indeed stabilize or shrink, for a while. But populations are not uniform. What of the 1 per cent, say, who are members of a minority with different values? If that minority has a growth rate of 5 per cent per year, then in ninety-five years they will be the majority, and in one thousand years their population will have grown by a factor of 1,500,000,000,000,000,000,000, if resource limits or genocide haven't intervened. Note that the Hutterites of North America, a reasonably wealthy religious group viewing fertility control as a sin and high fertility as a blessing, have managed an average of ten children per woman. Given enough time, exponential growth of even the smallest population can consume all the resources in reach. The right to reproduce is often regarded as basic, as illustrated by the outrage at reports of forced abortion in the People's Republic of China. The Hutterites and many others regard it as part of their freedom of religion. But what happens when parents have more children than they can support—does redistribution solve the problem? If reproduction is not forcibly suppressed, and if resources are forcibly and repeatedly redistributed so that each human being has a roughly equal share, then each person's share will steadily shrink. Even given the most optimistic assumptions regarding available resources, with a policy of resource redistribution and unlimited reproduction, the amount per person would eventually be insufficient to sustain life. This policy must be avoided, because if it is followed, it will kill everyone. As soon as we grant that any entity is entitled to certain rights—whether that entity be a human child, an animal, or some future artificial intelligence—the question arises of who is responsible for providing resources to support it when it can't do so for itself. The above argument indicates that a policy of coercion by some central power to compel the entire population to support an exponentially exploding population of these individuals would lead directly to disaster. Ultimately, this responsibility must rest with the entities' initiator: the designer of the artificial intelligence, the owner of the pet, the parents of the child. No new technology can magically remove the limits imposed by natural law, and thereby lift the burden of human responsibility.

Every time a technology solves a problem, it creates new problems. This doesn't mean that the change is neutral, or for the worse, of course. The Salk and Sabin vaccines for polio virtually destroyed the iron-lung industry, and the pocket calculator virtually destroyed the slide-rule industry, but these advances were worth the price of some economic adjustment. Molecular manufacturing and nanotechnology will bring far greater changes, placing far greater strains on our ability to adapt. We shouldn't be surprised when basically beneficial applications make someone miserable. Our lives are largely centered around problems. If we can solve many of these problems, the centers of our lives will shift, creating fresh problems. This section sketches some of the issues of change and adaptation more to raise questions than to offer solutions.

Molecular manufacturing offers the possibility of drastic change, a change in the means of production more fundamental than the introduction of industry, or of agriculture. Our economic and social structures have evolved around assumptions that will no longer be valid. How will we handle the changes in the way we work and live? Nanotechnology will have wide-ranging impact in many areas, including economic, industrial, and social patterns. What do historical patterns in similar circumstances tell us about the future? Any powerful technology with broad applications revolutionizes lives, and nanotechnology will be no exception. Depending on one's point of view, this may sound exciting or it may sound disturbing, but it most certainly does not sound comfortable. In comparison to many projections of the twenty-first century, though, nanotechnology may lead to comparatively comfortable change. The changes most often projected—for a future not including nanotechnology—have been ecological disaster, resource shortages, economic collapse, and a slide back into misery. The rise of nanotechnology will offer an alternative—green wealth—but that alternative will bring great changes from the patterns of recent decades. Times of rapid technological change are disconcerting. For most of humanity's existence, people lived in a stable pattern. They learned to live as their parents had lived—by hunting and gathering, later by farming—and changes were small and gradual. A knowledge of the past was a reliable guide to the future. Sudden changes, when they did occur, were apt to be ruinous: invasions or natural disasters. These sudden changes were fought or repaired or survived as best one could. Making major changes by choice was rare, and radical innovations were generally for the worse: the old ways at least ensured the ancestors' survival, the new might not. This made cultures conservative. It is only natural that there be efforts to resist change, but before undertaking such an effort, it makes sense to examine the record of what works and what doesn't. The only examples of successful change fighters have been communities that have created and maintained barricades to isolate themselves from the outside world socially, culturally, and technologically. For the two centuries before 1854, Japan turned its back on the outside world, following a deliberate policy of seclusion. The leaders of Albania restricted contacts for many years; only recently have they started to open up. Isolation attempts have worked better on a smaller scale, when participation is voluntary rather than decreed by government. Today, within the Hawaiian island chain, the tiny, privately owned island of Niihau, sixteen miles long and six miles wide, is deliberately kept as a preserve of the nineteenth-century Hawaiian lifestyle. Over two hundred full-blooded Hawaiians there speak the Hawaiian language and use no telephones, plumbing, television, and no electricity (except in the school). The Amish of Pennsylvania have no surrounding ocean to help maintain their isolation, but rely instead on tight social, religious, and technological rules aimed at keeping external technology and culture out, and themselves grouped in; those who leave the fold are excluded. On a national scale, attempts to take only one part of the package—whether social or technological—haven't done well at all. For decades, the Soviet Union and the Eastern bloc nations welcomed Western technology but attempted tight restrictions on the passage of people, ideas, and goods. Yet illegal music, thoughts, literature, and other knowledge still crept in—as they do into the Islamic countries. Fighting technological change in society at large has had little success, where that change gave some large group what it wanted. The most famous fighters of technological change—the Luddites—were unsuccessful. They smashed "automated" textile machinery that was replacing old hand looms during the early industrial revolution in England, but people wanted affordable clothing, and smashing equipment in one place just moved the business elsewhere. Change has sometimes been postponed, as when a later group, under the banner of "Captain Swing," smashed hundreds of threshing machines in a wide area of southern England in 1830. They succeeded in keeping the old, labor-intensive ways of harvesting for over a generation.

In previous centuries, when the world was less tightly connected by international trade, communications, and transportation, delays of years and even decades could be enforced through violence or legal maneuvers such as tariffs, trade barriers, regulations, or outright banning. Attempting to stop or postpone change is less successful today, when technology moves internationally almost as easily as people do—and human travel is so easy that 25 million people cross the Atlantic each year. Change fighters find that the problems they create mount with time. Products made using the old, high-cost techniques are uncompetitive. There is no way to bring back the "old jobs": they no longer make sense. But old habits die hard, and these same responses to the prospect of technological change continue today—ignoring it, denying it, and opposing it. Societies that have fought change, as Britain did, have fallen behind in a cloud of coal smoke. Why did the Luddites respond violently? Perhaps their response can be attributed to three factors: First, the change in their lives was sudden and radical; second, it affected a large group of people at one time, in one area; and third, in a world unprepared for rapid technological change, there was no safety net to catch the unemployed. While local economies might have been able to absorb a trickle of hungry laid-off workers, they lacked the size and diversity needed to offer other employment options quickly to large numbers of unemployed. In the twentieth century, however, societies have of necessity become somewhat better adapted to change. This has been a matter of necessity, because sluggish communities soon fall behind. In the ancient days of peasant stability, there was no need for institutions like Consumer Reports to study and rate new products, or regulators like the Environmental Protection Agency to watch over new hazards. We developed the needs, and we developed the institutions. These mechanisms represent important adaptions, not so much to the technologies of the twentieth century, but to the increasing change in technology during the twentieth century. There is great room for improvement, but they can perhaps provide a basis for adapting to the next century as well. Even with the best of institutions to cushion shocks and discourage abuse, there will be problems. The very act of solving problems of production—of increasing wealth—will create problems of economic change.

Over centuries, the trend has seemed to be toward centralization, beginning with the rise of factories and industrial towns. What drove these developments was the high cost of machinery and plant operations, the need to be near power sources, the impracticality of transportation among many small, dispersed sites, and the need for face-to-face communication. Beginning with the first industrial revolution, factories employed large numbers of people in one place, leading to overcrowding and making local economies dependent on one industry and sometimes on a single company. Costly equipment necessitated central locations for textile production, rather than the cottage industries where a lone woman could earn a livelihood carding wool and creating thread on a spinning wheel (providing the origin of the term *spinster*). By the 1930s, the belief in the virtures of centralization and central planning—the supposed efficiencies and economies of scale—led to nationwide or continentwide experiments in centralization. But over the last decade, these large-scale experiments have been dismantled, from Britain's privatization of nationalized utilities to the beginning of a return to the market system in Eastern European countries. Because the old limits on transportation, energy sources, and communication have fallen, business is now decentralizing. Between 1981 and 1986, the Forbes 500 companies cut their employees by 1.8 million. But during those same years, total civilian jobs went up by 9.2 million. Start-up companies created 14 million jobs; small companies created another 4.5 million. Telecommuting is booming, as are new businesses, independent professionals, and cottage industries. We've also seen the resurgence of small, but highly diverse stores: gourmet food shops, specialty ethnic shops; tea and coffee purveyors, organic and health food stores, bakeries, yogurt shops, gourmet ice-cream stores, convenience stores offering twenty-four-hour access, shops selling packaged food plus snacks. These stores epitomize something fundamental: At some point, what we want is not a standard good at an ever cheaper price, but special things customized to meet our own individual tastes or needs. The trend for advanced technologies seems to be leading away from centralization. Will nanotechnology counter or accelerate this trend? By reducing the cost of equipment, by reducing the need for large numbers of people to work on one product, and bringing greater ability to produce the customized goods that people want, nanotechnology will probably continue the twentieth-century trend toward decentralization. The results, though, will be disruptive to existing businesses. The computer industry perhaps provides a clue to what might happen as costs are lowered by nanotechnology. The computer-software industry is characterized by the garage-shop start-up. When your equipment is cheap—inexpensive PCs built around low-cost chips—and you can make a product by throwing in some ingenuity and human labor, it's possible to start a new industry on a shoestring. In 1900, when cars were simple, there were many car manufacturers. By the 1980s, if you weren't an industrial giant like General Motors or Ford, Honda or Nissan, you had to be John De Lorean to even get a shot at acquiring the capital to play in the business. If molecular manufacturing can slash the capital costs for producing cars or other plant-intensive equipment, we will see the equivalent of garage-shop businesses springing up to offer new products, and hiring workers away from the industrial giants of today just as the personal computer has destroyed the dominance of the mainframe.

The U.S.A. dream is to be an entrepreneur, and the technological trends of the twentieth century point in that direction. Nanotechnology probably continues it. In one area, however, the late twentieth-century trend has been toward uniformity. The nations of Western Europe are in the process of uniting under one set of economic rules, and parts of Eastern Europe are anxious to join them. More and more supranational and transnational organizations knit the world together. The growth of trade has motivated economic integration. Molecular manufacturing will work against this trend as well, permitting radical decentralization in economic terms. This will help groups that wish to step aside from the stream of change, enabling them to be more independent of the turbulent outside world, picking and choosing what technologies they use. But it will also help groups that wish to free themselves from the constraints of the international community. Economic sanctions will have little force against countries that need no imports or exports to maintain a high standard of living. And export restrictions will likewise do little to hamper a military buildup. By weakening the ties of trade, molecular manufacturing threatens to weaken the glue that holds nations together. We need that glue, though, to deal with the arms control issues raised by molecular manufacturing itself. This problem, caused by the potential for decentralization, may loom large in the coming years.

Lester Milbrath, professor of sociology and political science, observes,

"Nanotechnologies will create the problem of how to meaningfully and sustainably occupy the time of people who need not perform much work in order to have a sufficiency of life's goods. Our society has never faced this problem before, and it is not clear what social restructuring will be required to have a good society in those circumstances. We face much deep social learning." The world has had little experience with what anthropologists call "abundance economies." The native American tribes of the Pacific Northwest were one of those rarities. Ruth Benedict, in her classic book *Patterns of Culture*, wrote, "Their civilization was built upon an ample supply of goods, inexhaustible, and obtained without excessive expenditure of labor." The Kwakiutls became famous for their "potlatches": contests in which they sought to shame their rivals by heaping more gifts upon them than they could ever return. The potlatches would often be a year in preparation, last for days, and occasionally involve destruction of entire buildings. It was certainly a colorful form of keeping up with the Joneses. What will motivate us, once we have achieved an abundance economy? What will we regard as worthwhile goals to pursue? Increased knowledge, new art, improved philosophy, eliminating human and planetary ills? Will we find ourselves creating a better, wiser world, or sunk in boredom and jaded now that we have all and want nothing? If boredom gets out of hand, the lively spectacle of wealthy donors seeking to outdo each other to endow the arts, aid the poor, and do other good deeds for the sake of prestige would be welcome. What will happen as life spans continue to lengthen and the time needed to make a living decreases? Even today, there are people who, when confronted with the prospect of a significantly longer life span, exclaim that they couldn't imagine what they would do with all that time. This response can be hard to understand, when it would take a thousand years to walk all the world's roads, more thousands of years to read all the world's books, and another ten thousand years to have a dinner conversation with each of the world's people—but tastes differ, and even a few decades of bad television might make anyone long for the peace of the grave.

A major concern, and certainly the single area of greatest upheaval, is employment (which may become hard to distinguish from leisure). Once, people had little choice of employment. To keep a full belly, most had to work at the only job available: peasant farming. Eventually, people will have a complete choice of employment: they will be able to keep a full belly and a wealthy lifestyle while doing whatever they please. Today, we are about halfway between those extremes. In advanced economies, many different jobs are deemed useful enough that other people will offer an adequate income in exchange for the result. Some people can make a living doing something they enjoy—is this work, or leisure? The impact of nanotechnology on patterns of employment will depend on when it arrives. Current demographics show a shrinking supply of young people entering the work force. Agriculture, the assembly line, and entry level service jobs are experiencing a labor shortage, and no relief is in sight. If these trends continue, nanotechnology may show up in the midst of a shortage of labor. If it arrives late enough, it may compete with industries that are already nearing full automation; "job displacement" may mean replacing an industrial robot with a nanomachine. Employment patterns have shifted radically in the past. One hundred and fifty years ago, the United States was an agricultural nation—69 per cent of all people worked the land and a growing percentage worked in industry doing things like building steam locomotives for Baldwin Locomotives Works or tanning leather for the giant Central Leather monopoly. By the early twentieth century, agriculture was waning in numbers but increasing in productivity; most people worked in industry, and the tiny information and service sector was beginning to grow. Today, the picture has reversed: 69 per cent of employed Americans work in information or service jobs, only 28 per cent work in industrial production, and 3 per cent in agriculture. This tiny fraction feeds the other 97 per cent of Americans, exports hugely to other countries, and receives subsidies and price support payments to stop them from growing even more food. Manufacturing, even without nanotechnology, seems to be heading toward a similar condition. With an ever-declining percentage of our population working in manufacturing, we have as everyday products things that were once available only to kings and the high nobility. Yet owning multiple suits of clothes, having personal portraits of ourselves and family members, having music upon our command, having a personal bedroom, and having a coach awaiting our need—these are now regarded as being among the bare necessities of life. It may be possible to adjust to even greater wealth with even less required labor, but the adjustment will surely cause problems. In a world in which nanotechnology reduces the need for workers in agriculture and manufacturing still further, the question will be asked, "What jobs are left for people to do once food, clothing, and shelter are very inexpensive?" Again, the twentieth century provides some guidelines. As technology has reduced costs by efficiently producing many units of an identical item, people have begun to demand customization to meet individual needs or preferences. As a result, there are ever more jobs in producing custom goods. Today, semi-custom goods that try to help us meet our needs or express our taste abound: designer linens, ready-to-wear fashions, cosmetics, cars, trucks, recreational vehicles, furniture, carpeting, shoes, televisions, toys, sports equipment, washing machines, microwave ovens, food processors, bread bakers, pasta makers, home computers, telephones, answering machines—are all available in large and ever-changing variety.

Just as varied is the fabulous wealth and diversity of information produced in the twentieth century. Information products are a large factor in the economy: Americans buy 2.5 billion books, 6 billion magazines, and 20 billion newspapers each year. In recent years, new magazines have been invented and launched at the rate of one every business day of the year. A visit to a well-stocked magazine rack shows only a hint of the wealth of highly specialized publications, each one focused on a specialized interest or attitude: hotdog skiing, low-fat gourmet cooking, travel in Arizona, a magazine for people with a home office and a computer, and finely tuned magazines on health, leisure, psychology, science, politics, movie stars and rock stars, music, hunting, fishing, games, art, fashion, beauty, antiques, computers, cars, guns, wrestling. Motion pictures, which started as a flock of independent production companies and then consolidated into the great studios of the 1930s, have since followed the decentralization and diversification trends of recent years. Now an expanding range of film entertainment comes via network TV, cable channels, private networks, videotapes, music videos. Independent producers are aided by the technology innovations of cable, direct broadcast satellites, videotape technology, laser disks, videocameras.

The arts have burgeoned, with the general public as the new patron of the arts. Any artist or art form that could find and satisfy a market boomed in the twentieth century. Not just the traditional arts of actors, writers, musicians, and painters, but all forms of "domestic" artistry have grown to unprecedented levels: landscape and interior design, fashion design, cosmetics, hairstyling, architecture, bridal consulting. Providing for these demands are some of the "service and information" jobs created in the late twentieth century. "Service" jobs include many ways of helping other people: from nursing to computer repairs to sales. In "information" jobs, projected to have the fastest percentage growth over the next decade, people find, evaluate, analyze, and create information. A magazine columnist or TV news producer obviously has an "information" job. But so do programmers, paralegals, lawyers, accountants, financial analysts, credit counselors, psychologists, librarians, managers, engineers, biologists, travel agents, and teachers. "Increasingly," states *Forbes* magazine, "people are no longer laborers; they are educated professionals who carry their most important work tools in their heads. Dismissing them from their jobs, cutting them off from their places of employment may hurt them emotionally and financially. But it doesn't separate them from their vocation in the same way that pushing a farmer off his freshly seeded land does. For centuries workers were more dependent on a particular physical setting than they are now. Modern occupations generally give their practitioners more independence—and greater mobility—than did those of yesteryear." These human skills that people carry with them will continue to be valued: managing complexity, providing creativity, customizing things for other people, helping people deal with problems, providing old services in new contexts, teaching, entertaining, and making decisions. A reasonable guess would be that many of the service and information industries of the twentieth century will continue to evolve and exist in a world with nanotechnology. What is harder to imagine would be what new industries will come into being once we have new capabilities and lower costs. Along with the old economic law of supply and demand is another governing factor: price elasticity effects. People's desire for something is "elastic": it expands or contracts when the cost of something valuable goes down or up. If the price of a flight to Europe is five hundred dollars, more people will take a European vacation than if the price is five thousand dollars. When you had to hire a highly trained mathematician to do equations, calculation was slow and expensive. People didn't do much of it unless they absolutely had to. Today, computers make calculation cheap and automatic. So now businesses do sophisticated financial modeling, chemists design protein molecules, students calculate orbital trajectories for spaceships, children play video games, moviemakers do ever more amazing special effects, and the cartoon—virtually extinct because of high labor costs—has returned to movie theaters, all because computers permit cheap calculation. Nanotechnology will offer new, affordable capabilities to these and other people. Today, it's as hard to predict what new industries will be invented as it would have been for the creators of the ENIAC computer to have predicted cheap, handheld game computers for children.

So rather than producing drastic unemployment, nanotechnology seems likely to continue the trend already seen today, away from jobs that can be automated and into jobs where the human perspective is vital. But the true possibilities are, as always in the modern world, beyond predicting. Major shifts in demographics always cause disruptions. Even when we know they are coming, we never prepare for them. Our plans are based on expectations of what will happen. If things don't go as expected, we find that we have "malinvested." Houston real estate was valuable and looked to become even more so when times were good for the oil business there; when the fortunes of the oil business changed, Houston real estate was found to have been overbuilt, overpriced, and many millions of dollars were lost. Lengthening life spans push people toward taking a longer-term perspective, but rapid rates of change force a shorter-term perspective in investments. Turbulence in technology and in governmental monetary policy have already shortened time horizons. Businesspeople once routinely built plants with a thirty-year useful life. Today, the rate of change is too fast, and uncertainty regarding inflation and potential changes in tax laws is too great for such investments to make sense. Faster change will shrink time horizons further. Governments have taken on themselves the burden of looking a lifetime ahead, and the Social Security Administration is in for some rough times. When Otto von Bismarck, Germany's Iron Chancellor, came up with the notion of a guaranteed old age pension, it was a cynically clever and low-cost way to gain popular goodwill. So few people lived to age sixty-five that the amounts paid out in pensions were a pittance.

After watching the German experiment for a handful of years, other governments began following suit. None of them expected a world like ours where a baby girl born in the United States today has an average life expectancy of 78.4 years—double that of Bismarck's time—and even this estimate is based on the faulty assumption that her medical care will be no better than her great-grandmother's was. At present, the Social Security Administration has two models: one they call "positive" and one they call "negative." In the "positive" model, people work like dogs until old age, retire, and promptly die—presumably before they've had a chance to collect substantial social security or medical benefits. In the "negative" model, people retire early, develop illnesses that require medical intervention, and then live a long time making doctor visits and hospital stays during those years. Plans based on these models deserve to be disrupted. A better, more realistic scenario would have people living and able to support themselves for a long time, with illnesses that can be handled easily and inexpensively. Present social security benefits are enough to provide a certain standard of living—food, housing, transportation, and so forth. In a future of great material wealth, these benefits will be easy to provide, and present projections of economic woe resulting from an aging population will seem quaint.

Back in the seventies, author Alvin Toffler brought out a book called *Future Shock*, describing how disturbing rapid change is for people. The book was a best-seller, but how much actual future shock has been seen in the past decade? Most people seem to have come through the last two decades pretty much all right, not in a state of shock at all. Rather than being shocked by technology, they are instead annoyed about pollution and traffic. Does this mean Toffler was wrong in predicting future shock? It's true that technology has been advancing rapidly in many areas over the past twenty years. But consider the average person's home life: How much of this rapid technological advance has shown up there? A great deal, yet most of it is hidden, unlike the earlier part of the century, where obvious change was the norm. Electric lights and appliances, automobiles, telephones, airplanes, radio, and television affected almost everyone's private life. One person's life could span the time from horse-and-buggy travel to watching the Moon landings on television. In contrast, the past twenty years have seen new technologies move more quietly into the home. The VCR and microwave oven don't seem nearly as revolutionary as earlier inventions. Telephone answering machines are useful but haven't caused major changes in lifestyles. Fax machines are handy, but they're much like having very fast mail, and as this is written, fax machines aren't yet in most homes. So it's not surprising that the average person has felt little future shock lately. New medicines taken as pills—which may be radically improved—look just like the earlier pills. The computerized bills that come in the mail aren't any more exciting to pay than the old human-prepared bills. This situation is unlikely to last. How much longer can technology advance so rapidly in so many fields without major effects on our lifestyles? There's been a respite from future shock in the last three decades; people have had a chance to catch their breath. When nanotechnology arrives, will future shock arrive with it? Some segments of society today are already getting practice in dealing with rapid technological advance. Those getting the most vigorous workout are in the computer field, where a machine two years old is regarded as obsolete, and software must be updated every few months to keep abreast of the new developments. But has this terrific rate of progress been dizzying or overwhelming? Not for the consumer—on the contrary, computers have become easier to use. In the 1960s, the New Math that was introduced into American grade schools and junior high schools included extensive study of arithmetic using numbers written in something other than the familiar base 10. This was to prepare the "Adults of Tomorrow" for "The Computer Age" in which we would all be writing assembly language computer programs in binary (base 2) code. But customers now purchase software rather than write it themselves—they need never deal with computer languages at all, much less a primitive assembly language. The rapid increase of computer speed has helped make computers easier to use. This progression has occurred many times before: Cars started off with external hand cranks, then advanced to starters you could yank from the comfort of the driver's seat; now starters perform invisibly when you turn the key in the ignition.

This pattern will surely continue. First, some people will adapt to the technology, but in the long run the technology will adapt to us. The more flexible and powerful the technology, the more easily it will adapt. Seen from a distance, seemingly trivial patterns of adaptation form part of a larger process that has marked the last century: the Western world has begun to invent mechanisms to handle a world of persistent change. Our mechanisms are by no means perfect or painless, as any unemployed person can testify. Employment agencies and headhunters for job seekers; unemployment and severance packages to ease job transitions; on-the-job training, continuing education, retraining, specialized seminars to update professional skills, professional associations, networking, community resources centers, government training programs, and volunteer agencies are just a few of the inventions dealing with change and transition. Consumer information services, regulatory agencies, and environmental organizations are others. The most effective will endure. More options will continue to be invented. intellectual exercise of Feynman's had its obvious seed in the then recent (1953) discovery of the DNA double helix by Watson and Crick. If the genome has such a small size, Feynman thought, given the awesome amount of information it encodes, is there anything preventing scientists and engineers from emulating this natural feat?

Let us keep in mind such inheritance. Biology thus gave birth to the whole field of nanotechnologies. To go by the empirical criterion, Feynman's call was that of a pioneer, it went unheeded for a generation.

After Heinrich Rohrer and Gerd Binnig devised in 1981 a scanning tunneling microscope, a collective leap forward ensued. About that time, George B. Whitesides initiated a series of studies which showed the kind of contributions chemists might make to nanosystems. The nanotechnology bandwagon started rolling in the 1990s. It was pulled by two horses. There was Moore's Law. Engineers and scientists had for decades looked at the horizon of computers foregoing silicon chips in favor of molecular devices. And there was the demographic factor. The steady increase in the number of scientists, which continued well into the 1980s, was unmatched by a corresponding increase in public money. Accordingly, young scientists as a group partitioned themselves between industrial and pharmaceutical work.

New Techniques on Old Steps

But let us return to the attitude of chemists who discard nanotechnologies as a combination of old, well-known chemistry and a public relations ploy to raise money from industry and from granting agencies. Could it be true that the field of nanotechnologies amounts only to renaming a subset of well-established chemical knowledge? Does it lack originality and impetus? Are we dealing with old wine, which time may have started to sour and to make unpalatable and thus needed a new label? Conversely, is this new wine whose novel taste suggests brand-new flasks? Consider colloidal gold as a test case. Gold nanoparticles have been the subject of considerable attention over the ages. A back-of-the-envelope calculation shows that a small nanocrystal of one nanometer diameter will have as many as 30% of its atoms on the surface while a larger nanocrystal of 10 nm (ca. 1000 atoms) will have about 15% of its atoms on the surface. Metal particles with dimensions in the nanometer scale thus display unusual properties. The mean free path of an electron in a metal at room temperature is in the range 10-100 nm. As a metallic particle shrinks to that dimension, unusual effects might be observed. Indeed, gold nanoparticles of diameters about 100 nm or less appear red, not gold, when suspended in transparent media.

They enjoy an interesting history dating back to the pioneering work of Faraday on the synthesis of gold hydrosols, gold nanoparticles dissolved in water. As one may recall, early in 1856 Michael Faraday (1791-1867) became interested in the interaction of gold with light. Gold leafs show different colors in transmitted light, green, blue, and purple, than they do in reflected light. This observation intrigued Faraday. Together with his friend, Warren de la Rue, he prepared gold films. He started preparation of gold precipitates from reducing with ferrous sulfate a solution of auric salts, made presumably by attack of the metal with *aqua regia*. In so doing, Faraday discovered colloidal gold. This was the subject of intense experimenting and thought. Between the onset of this work, February 2, and its completion at the end of the same year 1856, Faraday wrote in his diary no fewer than 300 manuscript pages (1,160 numbered entries) on this topic. He submitted an article to the Royal Society on November 15, it appeared on February 15, 1857. In general and since Faraday, metal sols possess fascinating colors and have long been used as dyes.

Smaller gold nanoparticles of diameters about 3 nm are no longer 'noble' and unreactive, but can catalyze chemical reactions. Consider nanocatalysts comprised of gold atom clusters, Au_8, Au_4, Au_3Sr, adsorbed at an F center of a MgO (100) plane. The optimal adsorption geometries of the O_2 molecule to these nanocatalysts differ markedly. Another example is from the temperature-programmed reaction spectra of the different products of the polymerization of acetylene on small supported, monodispersed palladium clusters. Striking atom-by-atom size-dependent reactivities and selectivities are observed. Only the three reaction products C_6H_6 (benzene), C_4H_8 (butene), and C_4H_6 (butadiene) form. Remarkably no C_3H_n, C_5H_n, and C_8H_n are detected, indicating the absence of C-C bond scission, as already observed on palladium single crystals and on palladium particles. Up to Pd_3, only benzene formation is catalyzed, reflecting a high selectivity for cyclotrimerization of acetylene. Pd_n clusters ($4 < n < 6$) reveal a second reaction channel by catalyzing in addition the formation of butadiene. The third reaction product, butene, desorbing at a rather low temperature of 200 K, is clearly observed for Pd_8. For this cluster size the abundance of the three reaction products is similar. For even larger clusters ($13 < n < 30$) benzene formation increases with cluster size, whereas the conversion of acetylene to butene reaches a maximum for Pd_{20}. Note that Pd_{30} selectively suppresses the formation of butadiene (see for instance the Nanocatalysis chapter in the book by Rao, Müller, and Cheetham).

What this example of metallic nanoparticles shows, which many of us have always suspected to be true but which conventional wisdom and standard practice – why bother reading any scientific publication older than say five years? – have both condemned summarily, is that casks of old wine connect with containers of the precious liquid from the most recent harvest. New science can be at its most innovative when it grafts itself onto its historical roots. Prediction starts with retrodiction, a point to which I will return.

Who is most responsible for inflating the nanobubble? Are scientists, journalists, or venture capitalists the guiltiest? Whatever the answer, a host of start-up companies have seen the light of day, and have issued shares. The Yellow Pages will have to make room for quite a few names with the nano- prefix: Nanometrics, Nanogen, Nanologic, Nanopierce, Nanoproprietary, Nanosys, Nanotechnology Development Corporation are just a few. Some universities have invested heavily into setting-up industrial parks specialized in nanotechnology. SUNY Albany and the University of South Carolina are two examples. The promise is there, because some of the devices already made actually work. IBM was able to use a nanotechnique to increase by a factor 20 the amount of memory on a hard drive.

John Wolfe, who started the first nanotech venture-capital firm, compares it to the Internet circa 1993, before Netscape went public. Is it to say that this new bubble is doomed from the start and that it will burst, sooner or later? Assuredly. Is it to say that it will sink without leaving a trace? Assuredly not. Some companies will survive and be lasting. Already, one can differentiate between fly-by-night operations buoyed only by conning the gullible, and reputable firms such as Nanosys doing solid work, and worth investing into. Nanosys collaborates with well-established corporate giants such as Intel, Matsushita, and DuPont. Peugeot-Citroën and Ford inject sub-10 nm CeO_2 nanoparticles into diesel fuel before it is burnt in the engine. This results in an *in situ* catalyst that not only improves fuel efficiency but also greatly reduces the emission of particulates. More than a million automobiles will incorporate that system by 2005 already. The American Federal government is also among the investors in the Nano Rush. Washington is spending $ 3.7 billion on nanotech research over the next four years. To this writer, the key issue is representation. Social groups create self-images, they project somewhat different images to the public. Nanotechnology, as mentioned at the beginning of this essay, arose from one of the icons of the 20th century, the DNA double helix. It was successful, early on, in giving itself a similar and logo-like icon. IBM scientists created it and the advertising arm of the multinational corporation was ready to pounce and was quick to hammer it irreversibly into public consciousness. This is the STM image of a corral of 48 iron atoms placed in a circle of 7.3 nm radius. This particular image has become emblematic of nanotechnologies, at least for the time being. With it, scientists exhibited to the public, not without intense pride, an atom-made fence. What a feat! But this construction can also be seen as an unselfconscious picture of the proverbial ivory tower. In yet another image, nanotechnology proudly depicted in 1990 on the cover of *Nature* its archetype with an STM image of DNA.

Another graphic depiction of nanotechnologies brings up my next point, representation as re-presentation. The cover page in issue number 24 of *Angewandte Chemie International Edition in English,* volume 37 for 1998, was adorned with a schematic representation of the Mo_{132} cluster. Its similarity to Kepler's early model of the universe, featuring the ubiquity of the icosahedron, was emphasized in the illustration.

Thus, nanotechnology in the time of the Millennium rediscovered the microcosm-macrocosm correspondence prevailing in medieval metaphysics. With some modification, needless to say. Humanism is out: a metallic cluster as the nanocosm replaces man as the microcosm. Just as for medieval theologians, the stress is on the unity of nature, on the uniformity of physical law throughout the universe. But physical science has replaced God in Western thought.

One can interpret also the bow to Kepler as one of the numerous reincarnations of the Platonic archetype of the five regular solids, which has marked the history of chemistry in the braudelian *longue durée*: Kepler and the snowflake, the constituent particles of crystalline solids as proposed by René-Just Haüy, molecular shape according to André-Marie Ampère, Achille Le Bel's and Jacobus Henricus van't Hoff's tetrahedral carbon, Alfred Werner's complexes, Aaron Klug's icosahedral viruses and, closer yet to us, Richard Smalley's and Harold Kroto's buckminsterfullerene. Thus, nanochemists gave themselves a self-image of continuity within the geometric, Pythagorean tradition.

Let me sum up with representation as re-presentation. I have been at pains in this review to point to all the aspects of nanotechnology that predated it. Nanoscientists have moved existing pieces to create new patterns. There is nothing wrong with this efficient heuristics. Discovery is not the only path to knowledge, fame, and riches. But this raises an interesting question in philosophy and historiography, which will be my final point. Science rewrites its own history, not only as history but as science fodder too. Which should not be construed in the weak sense of Whig history-telling as indulged in by scientists. While it is very much true that scientists, in a Whig perspective, tend to see the past as heralding the present, their purpose is not the mundane one of setting up an artificial history, approximating to a linear genealogy. The nanotechnologies paradigm shows a considerably more interesting attitude. Science reconfigures past events and developments in order to refocus itself into new directions. Such a reshuffling of the past, which of course may be deeply upsetting to professional historians who see their work voided and having to be restarted anew, the undersigned included, is an essential part of the dynamics of science. And it shows historians of chemistry that there is life after Partington.

Building large objects is basic to solving problems of housing and transportation. Smart materials can help. Today, buildings are expensive to construct, expensive to replace, and expensive to make fireproof, tornado-proof, earthquake-proof, and so forth. Making buildings tall is expensive; making walls soundproof is expensive; building underground is expensive. Efforts to relieve city congestion often founder on the high cost of building subways, which can amount to hundreds of millions of dollars per mile. Building codes and politics permitting, nanotechnology will make possible revolutions in the construction of buildings. Superior materials will make it easy to construct tall (or deep) buildings to free up land, and strong buildings that can ride out the greatest earthquake without harm. Buildings can be made so energy-efficient and so good at using the solar energy falling on them that most are net energy producers. What is more, smart materials can make it easy to build and modify complex structures, such as walls full of windows, wiring, plumbing, data networks, and the like. For a concrete example that shows the principle, let's picture what smart pipes could be like. Let's say that you want to install a fold-down sink in the corner of your bedroom. The new materials make fold-down sinks practical, and in a house made of advanced smart materials, just sticking one on the wall would be enough—the plumbing would rearrange itself. But this is an old, pre-breakthrough house, so the sink is a retrofit. To do this home-handiwork project, you buy several boxes full of inexpensive tubing, T-joints, valves, and fixtures in a variety of sizes, all as light as wood veneer and feeling like soft rubber. The biggest practical problem will be to make a hole from an existing water pipe and drainpipe to where you want the sink. Molecular manufacturing can provide excellent power tools to make the holes, and smart paint and plaster to cover them again, but the details depend on how your house is built. The smart plumbing system does help, of course. If you want to run the drain line through the attic, built-in pumps will make sure that the water flows properly. The flexibility of the pipes makes it much easier to run them around curves and corners. Low-cost power makes it practical for the sink to have a flow-through water heater, so you only need to run a cold-water pipe to have both hot and cold water. All the parts go together as easily as a child's blocks, and seem about as flimsy and likely to leak. When you turn it on, though, the microscopic components of the pipes lock together and become as strong as steel. And smart plumbing doesn't leak. If your house were made of smart materials, like most of the housing in the Third World these days, life would have been easier. Using a special trowel, wall structures would be reworked like soft clay, doing their structural job all the while. Setting up a plumbing system from scratch with this stuff is easy, and hard to do wrong. Drinking water pipes won't connect to wastewater pipes, so drinking water can't be accidentally contaminated. Drains won't clog, because they can clean themselves better than a rotary steel blade ever could. If you run enough pipes from everything to everything else, built-in pumps will make sure that water flows in the right direction with adequate pressure. Smart plumbing is one example of a general pattern. Molecular manufacturing can eventually make complex products at low cost, and those complex products can be simpler to use than anything we have today, freeing our attention for other concerns. Buildings can become easy to make and easy to change. The basic conveniences of the modern world, and more, can be carried to the ends of the earth and installed by the people there to suit their tastes.

Worldwide food production has been outpacing population growth, yet hunger continues. In recent years, famine has often had political roots, as in Ethiopia where the rulers aim to starve opponents into submission. Such problems are beyond a simple technological solution. To avoid getting headaches, we'll also ignore the politics of farm price-support programs, which raise food prices while people are going hungry. All we can suggest here is a way to provide fresh food at lower cost with reduced environmental impact. For decades, futurists have predicted the coming of synthetic foods. Some sort of molecular-manufacturing process could doubtless make such things with the usual low costs, but this doesn't sound appetizing, so we'll ignore the idea. Most agriculture today is inefficient—an environmental disaster. Modern agriculture is famed for wasting water and polluting it with synthetic fertilizers, and for spreading herbicides and pesticides over the landscape. Yet the greatest environmental impact of agriculture is its sheer consumption of land. In the American East, ancient forests disappeared under the ax, in part to supply wood, in part to clear land. The prairies of the West disappeared under the plow. Around the world, this trend continues. The technology of the ax, the fire, and the plow is chiefly responsible for the destruction of rain forests today. A growing population will tend to turn every productive ecosystem into some sort of farmland or grazing land, if we let it. No technological fix can solve the long-term problem of population growth. Nonetheless, we can roll back the problem of the loss of land, yet increase food supplies. One approach is intensive greenhouse agriculture. Every kind of plant has its optimum growing conditions, and those conditions are far different from those found in most farmland during most of the year. Plants growing outdoors face insect pests, unless doused with pesticide, and low levels of nutrients, unless doused with fertilizer. In greenhouses patrolled by "nanoflyswatters" able to eliminate invading insects, plants would be protected from pests and could be provided with nutrients without contaminating groundwater or runoff. Most plants prefer higher humidity than most climates provide. Most plants prefer higher, more uniform temperatures than are typically found outdoors. What is more, plants thrive in high levels of carbon dioxide. Only greenhouses can provide pest protection, ample nutrients, humidity, warmth, and carbon dioxide all together and without reengineering the Earth. Taken together, these factors make a huge difference in agricultural productivity. Experiments with intensive greenhouse agriculture, performed by the Environmental Research Lab in Arizona, show that an area of 250 square meters—about the size of a tennis court—can raise enough food for a person, year in and year out. With molecular manufacturing to make inexpensive, reliable equipment, the intensive labor of intensive agriculture can be automated. With technology like the deployable "tents" and smart materials we have described, greenhouse construction can be inexpensive. Following the standard argument, with equipment costs, labor costs, materials costs, and so forth, all expected to be low, greenhouse-grown foods can be inexpensive. What does this mean for the environment? It means that the human race could feed itself with ordinary, naturally grown, pesticide-free foods while returning more than 90 per cent of today's agricultural land to wilds. With a generous five hundred square meters per person, the U.S. population would require only 3 per cent of present U.S. farm acreage, freeing 97 per cent for other uses, or for a gradual return to wilderness. When farmers are able to grow high-quality foodstuffs inexpensively, in a fraction of the room that they require today, they will find more demand for their land to be tended as a park or wilderness than as a cornfield. Farm journals can be expected to carry articles advising on techniques for rapid and esthetic restoration of forest and grassland, and on how best to accommodate the desires of the discriminating nature lover and conservationist. Even "unpopular" land will tend to become popular with people seeking solitude. The economics of assembler-based manufacturing will remove the incentive to make greenhouses cheap, ugly, and boxy; the only reason to build that way today is the high cost of building anything at all. And while today's greenhouses suffer from viral and fungal infestations, these could be eradicated from plants in the same way they would be from the human body, as will be described later. A problem faced by today's greenhouses—overheating—could be dealt with by using heat exchangers, thereby conserving the carefully balanced inside atmosphere. Finally, if it should turn out that a little bit of bad weather improves the taste of tomatoes, that, too, could be provided, since there would be no reason to be fanatical about sheer efficiency.

Today, telecommunications systems have sharply limited capacity and are expensive to expand. Molecular manufacturing will drop the price of the "boxes" in telecommunications systems—things such as switching systems, computers, telephones, and even the fabled videophone. Cables made of smart materials can make these devices easy to install and easy to connect together. Regulatory agencies willing, you might someday be able to buy inexpensive spools of material resembling kite string, and other spools of material resembling tape, then use them to join a world data network. Either kind of strand can configure its core into a good-quality optical fiber, with special provisions for going around bends. When rubbed together, pieces of string will fuse together, or fuse to a piece of tape. Pieces of tape do likewise. To hook up to the network, you run string or tape from your telephone or other data terminal to the nearest point that is already connected. If you live deep in a tropical rain forest, run a string to the village satellite link. These data-cable materials include amplifiers, nanocomputers, switching nodes, and the rest, and they come loaded with software that "knows" how to act to transmit data reliably. If you're worried that a line may break, run three in different directions. Even one line could carry far more data than all the channels in a television cable put together.

Getting around quickly requires vehicles and somewhere for them to travel. The old 1950s vision of private helicopters would be technically possible with inexpensive, high-quality manufacturing, cheap energy, and a bit of improvement in autopilots and air-traffic control—but will people really tolerate that much junk roaring across the sky? Fortunately, there is an alternative both to this and to building ever more roads.

Near the surface of the Earth, there is as much room underground as there is above it. This is usually ignored, because the room is full of dirt, rock, pressurized water, and the like. Digging is expensive. Digging long, deep tunnels is even more expensive. This expense, however, is mostly in the cost of equipment, materials, and energy. Tunneling machines are in wide use today, and molecular manufacturing can make them more efficient and less expensive. The energy to operate them will be no great problem, and smart materials can line tunnels as fast as they are dug, with little or no labor. Nanotechnology will open the low frontier. With a little care, the environmental impact of a deep tunnel can be trivial. Instead of solid rock far below the surface, there is rock with a sealed tunnel running through it. Nothing nearby need be disturbed. Tunnels avoid both the aesthetic impact of a sky full of noisy aircraft and the environmental impact of paving strips of landscape. This will make them less expensive than roads, and they can, if desired, be more common than roads in the developed world today. They will even permit faster transportation.

Japan and Germany are actively developing magnetic trains, like those in the Desert Rose Scenario. These avoid the limitations of steel wheels on steel rails by using magnetic forces to "fly" the train along a special track. Magnetic trains can reach aircraft speeds at ground level. On long runs through evacuated tunnels, they can reach spacecraft speeds, traveling global distances in an hour or so (less, if passengers are willing to tolerate substantial acceleration). Systems like this can give "taking the subway" a new meaning. Local transportation would be at fast automotive speeds, but long-distance transportation would be faster than the Concorde. With superconducting electrical systems, fast subways would be more energy efficient than today's slow mass transit.

For decades, people have proposed replacing automobiles with some form of mass-transportation system, and it seems that cost revolutions (including inexpensive tunneling) may finally make this practical. Before junking the car, though, it's worth seeing how it might be improved. Molecular manufacturing can make almost anything better. Automobiles can be made stronger and safer, lighter, higher performance, and higher efficiency, while getting excellent mileage and burning clean, inexpensive fuels, perhaps in fuel cells powering quiet electric motors. Using aerodynamic forces to hold the car to the road, there's no reason why a comfortable passenger car shouldn't be able to deliver uncomfortable, drag-racer acceleration. To imagine a cheap car built with molecular manufacturing, first imagine loading it with all the attractive features that you've ever heard proposed. This includes everything from today's self-adjusting seats and mirrors, excellent sound systems, and specially tuned steering and suspension systems, through automated navigation displays, emergency braking, and reliable super-duper airbags. Now, instead of just having the position of the seats, mirrors, and so forth adjust to a driver, as some cars do today, our smart-material car can also adjust its size, shape, and color, facing owners with choices such as, "What should our car look like for this occasion?" Those seeking an image of solid conservatism and wealth won't drive such cheap cars; they will risk their necks in a certified antique car, made from the traditional steel, paint, and rubber. If environmental regulations permit it, the car might even have a genuine gasoline-burning engine. The latter can no doubt be cleaned up by fancy nanotechnology-based emission-control systems.

Our transportion system today effectively ends in the upper atmosphere. Travel beyond still takes the form of "historic missions." There is no reason for this situation to continue for long, once molecular manufacturing becomes well established. The cost of spaceflight is high because spacecraft are huge, fragile things, made in such small numbers that they're almost hand-crafted. Molecular manufacturing will replace today's delicate monsters with rugged, mass-produced vehicles (which, with greater efficiency, needn't be so large). The vehicles will cost little, but the energy? Today, the energy cost of a ticket to orbit in an efficient vehicle would be less than one hundred dollars. Low cost vehicles and energy will drop the total cost to a fraction of this. We will know that spaceflight has become inexpensive when people see the Earth as just a small part of the world, and understand in their bones that space resources make continued exploitation of Earth's resources unnecessary. In the long run, efficient, clean, low-cost manufacturing can transform the way human beings affect the Earth by their presence. Even stay-at-home humans will be better able to heal the damage they have done.

REFERENCES

Attard, G.S., et al. 1997. *Science* 278: 838. Averback, R.S., J. Bernholc, and D.L. Nelson. 1991. *MRS Symposium Proceedings* Vol. 206.

B. Yurke, A.J. Turberfield, A.P. Mills, Jr., F.C. Simmel, J.L. Neumann, "A DNA-fuelled molecular machine made of DNA," *Nature* 406(10 August 2000): 605-608.

Ben-David, Joseph and Randall Collins, 1966. "Social Factors in the Origins of a New Science: the Case of Psychology." *American Sociological Review.* Vol. 31 pp 451-465.

Blumenthal, D. 1992. "Academic-Industry Relationships: Extent, Consequences, and Management," *The Journal of the American Medical Association*, 268: 3, December, pp. 3344 -3349.

Blumenthal, D.M., M. Gluck, K.S. Louis, M. Stoto, and D. Wise. 1986. "University-Industry Research Relationships in Biotechnology: Implications for the University," *Science* 232: pp. 1361-1366 (June 13).

BMBF. 1996. "Delphi-Bericht 1995 zur Entwicklung von Wissenschaft und Technik – Mini-Delphi." Bonn, author's translation.

Brown, J.S., and Paul Duguid. 2001. Don't count society out. *Societal Implications of Nanoscience and Nanotechnology.* National Science Foundation. Arlington, VA. March.

Callon, M.: 1986, 'Some Elements of a Sociology of Translation', in J. Law (ed.), *Power, Action and Belief: A New Sociology of Knowledge?*, Routledge, London, pp. 196-223.

Canguilhem, G.: 1952, *Machine et organisme' in La connaissance de la vie*, Hachette, Paris [quoted from the fourth edition Vrin, Paris, 1971].

Carroll, J.S.: 2001, 'Social Science Research Methods for Assessing Societal Implications of Nanotechnology', in: M.C. Roco, W.S. Bainbridge (eds.), *Societal Implications of Nanoscience and Nanotechnology*, National Science Foundation, pp. 188-192.

Carsley, J.E., R. Shaik, W.W. Milligan, and E.C. Aifantis. 1997. In *Chemistry and physics of nanostructures and related non-equilibrium materials*. ed. E. Ma, B. Fultz, R. Shull, J. Morral, and P. Nash. Warrendale, PA: TMS.

Chang, K. "Smaller Computer Chips Built Using DNA as Template." New York Times, November 21, 2003.

Chopra, N.G., R.J. Luyken, K. Cherrey, H. Crespi, M.L. Cohen, S.G. Louie, and A. Zettl. 1995. *Science* 269: 966.

Christensen, Clayton M. *The Innovator's Dilemma: When New Technologies Cause Great Firms to Fail*. New York: Harperbusiness, 2000.

Colbert, D.T. and R.E. Smalley. 1999. Fullerene nanotubes for molecular electronics. *Trends in Biotechnology*. Vol 17. 46-50.

Collier, C.P., E.W. Wong, M. Belohradsky, F.M. Raymo, J.F. Stoddart, P.J. Kuekes, R.S. Williams, and J.R. Heath. 1999. "Electronically Configurable Molecular-Based Logic Gates." *Science*, 285.

Constant, E.W., II: 1980, *The Origins of the Turbojet Revolution*, Johns Hopkins University Press, Baltimore.

Doering, Robert and Yoshio Nishi. 2000. "Limits of Integrated-Circuit Manufacturing." *Proceedings of the IEEE*. September.

Doi, M. and S.F. Edwards. 1986. *The Theory of Polymer Dynamics*. Oxford University Press (Clarendon), London-New York.

Donald, A.M. and A.H. Windle. 1992. Liquid crystalline polymers. *Cambridge Solid State Science Series*. Cambridge University Press.

Drell, S. D., A. D. Sofaaer, and G. D. Wilson, *The New Terror, Facing the Threat of Biological and Chemical Weapons*, Hoover Institution Press, Stanford University (1999).

Dresselhaus, M.S., G. Dresselhaus, and P. Eklund. 1996. *Science of fullerenes and carbon nanotubes*. San Diego: Academic Press. Gleiter, H. 1989. *Prog. Mater. Sci.* 33: 223.

Drexler K.E.: 1995, 'Introduction to nanotechnology', in: Krummenacker, M. & Lewis, J. (eds.), *Prospects in Nanotechology. Proceedings of the 1st general conference on nanotechnology: developments, applications, and opportunities, November 11-14, 1992, Palo-Alto*, John Wiley & Sons, New York, pp. 1-20.

Drexler, E., Phoenix, C.: 2004, 'Self-Replication in Nanotechnology: Feasible, Potentially Safe, and Unnecessary' (unpublished paper, in preparation).

Drexler, E.; Smalley, R.E.: 2003, 'Point-Counterpoint: Nanotechnology', *Chemical and Engineering News*, 81(48), 37-42.

Drexler, K. Eric. 1986. *Engines of Creation*. Garden City, N.Y.: Anchor Press/Doubleday.

Drexler, K.E.: 1981, 'Molecular Engineering—An Approach to the Development of General Capabilities for Molecular Manipulation', *Proceedings of the National Academy of Sciences of the United States of America-Physical Sciences*, 78, 5275-5278.

Dupuy, J.P. "Common Knowledge, Common Sense." Theory and Decision 27 (1989): 37-62.

Ellul, J. 1964. *The Technological Society*. New York: Alfred Knopf. (First published 1954).

Empedocles, S.; Bawendi, M. 1999. "Spectroscopy of single CdSe nanocrystallites," *Accounts of Chemical Research*, 32, 389-396.

Entingh, D., Dunn, A., Glassman, E., Wilson, J. E., Hogan, E. & Damstra, T. (1975) in *Handbook of Psychobiology*, eds. Gazzinga, M. S. & Blakemore, C. (Academic, New York), pp. 201-238.

Esant. 1999. "Economic and Social Aspects of Nanotechnology." First draft report of working group 6 of the Euroconferences on Nanoscience for Nanotechnology.

Estermann, M., L.B. McCusker, C. Baerlocher, A. Merrouche, and H. Kessler. 1991. *Nature* 331: 698.

Fedor N. Dzegilenko, Deepak Srivastava, Subhash Saini, "Simulations of carbon nanotube tip assisted mechano-chemical reactions on a diamond surface," *Nanotechnology* 9(December 1998): 325-330.

Feynman R. "There's Plenty of Room At the Bottom." Talk given on at the annual meeting of the American Physical Society at the California Institute of Technology, 1959.

Friedlander, S.K. 1998. Synthesis of nanoparticles and their agglomerates: aerosol reactors. In *R&D status and trends*, ed. Siegel et al.

Galison, P.: 1997, *Image and Logic: A Material Culture of Microphysics*, University of Chicago Press, Chicago.

George M. Whitesides, "The Once and Future Nanomachine," *Scientific American* 285(September 2001): 78-83.

Gleiter, H. 1990. *Progress in Materials Science* 33: 4. Guinier, A. 1938. *Nature* 142: 569; Preston, G.D. 1938. *Nature* 142: 570.

Gutte, B., Dannigen, M. & Wittschieber, E. (1979) *Nature (London)* 281, 650-655.

Hadjiconstantinou, N.G. 1999. Combining atomistic and continuum simulations of contact-line motion. *Physical Review E*. Vol. 59(2).2, 2475-2478.

Hadjipanayis, C.G. 1998. Nanostructured magnetic materials. In *R&D Status and Trends*, ed. Siegel et al.

Hamilton, S.: 2003, 'Biotechnology, SF and the Media', *Science Fiction Studies*, 30, 267-282.

Harold J. Morowitz, "A Model of Reproduction," *American Scientist* 47(1959): 261-263.

Haruta, M. 1997. *Catalyst surveys of Japan* 1: 61 and references therein.

Hayes, B., "The Square Root of NOT," *American Scientist*, July-August 1995, pp. 304-308.

Hayles, N.K. (ed.): 2004, *Nanoculture: Implications of the New Technoscience*, Bristol, UK: Intellect Books.

Hayles, N.K.: 2004, 'Connecting the Quantum Dots: Nanotechscience and Culture', in: N.K. Hayles (ed.), *Nanoculture: Implications of the New Technoscience*, Bristol, UK: Intellect Books, pp. 11-26.

Hayward, R.C.; Saville, D. A.; Aksay, I.A. 2000. "Electrophoretic assembly of colloidal crystals with optically tunable micropatterns", *Nature*, 404, 56-59.

Higgins, R.J. 1997. An economical process for manufacturing of nano-sized powders based on microemulsion-mediated synthesis. In *Proc. of the Joint NSF-NIST Conf. on Nanoparticles.*

Hiruma, K., M. Yazawa, T. Katsoyama, K. Ogawa, K. Haraguchi, M. Koguchi, and H. Kakibayashi. 1995. *J. Appl. Phys.* 77(2): 476.

Howland, H.: 1962, Structural, hydraulic and economic aspects of leaf venation and shape ', in: E.E. Bernard & M.R. Kare (eds.), *Biological prototypes and synthetic systems*, Plenum Press, New York, vol. 1, pp. 183-192.

Hoyningen-Huene, P.: 1993, *Reconstructing Scientific Revolutions: Thomas S. Kuhn's Philosophy of Science* (trans. by A. Levin), University of Chicago Press, Chicago.

Huang, J.Y., Y.K. Wu, and H.Q. Ye. 1996. *Acta Mater.* 44: 1211. Inoue, A. 1997. Private communication. Inturi, R.B., and Z. Szklavska-Smialowska. 1992. *Corrosion* 48: 398. Karch, J., R. Birringer, and H. Gleiter. 1987. *Nature* 330: 556.

Hubbell, J.A., and R. Langer. 1995. Tissue engineering. *Chem. Eng. News* (March 13): 42- 54.

Hughes, P. M., *Global Threats and Challenges: The Decades Ahead*, Lt. Gen. Patrick M. Hughes, Director of the Defense Intelligence Agency, in Statement for the Senate Committee on Intelligence, January 28, 1998.

Jameson, F.: 1982, 'Progress Versus Utopia; or, Can We Imagine the Future', *Science Fiction Studies*, 9, 147-158.

Janet Raloff, "Nano Hazards: Exposure to minute particles harms lungs, circulatory system," Science News Online, Week of March 19, 2005; Vol. 167, No. 12.

Jonas, H. The Imperative of Responsibility. In Search of an Ethics for the Technological Age. Chicago: University of Chicago Press, 1985.

Kline, R.: 2000, 'The Paradox of 'Engineering Science': A Cold War Debate about Education in the United States', *IEEE Technology and Society Magazine*, 19, 19-25.

Kline, R.; Pinch, T.: 1996, 'Users as Agents of Technological Change: The Social Construction of the Automobile in the Rural United States', *Technology and Culture*, 37, 763-95.

Kroes, P. & Meijers, A. "The Dual Nature of Technical Artifacts—Presentation of a New Research Programme." Techné 6, 2 (2002): 4-8.

Kuusi, O. and M. Meyer. 2001. "Technology Generalizations and Leitbilds – The Anticipation of Technological Opportunities." Manuscript submitted to *Technological Forecasting and Social Change*.

Kuusi, O., 1999. "Epistemic value analysis – a tool to evaluate technological options." Personal communication.

Kuusi, O.; Meyer, M.: 2002, 'Technological generalizations and leitbilder ' the anticipation of technological opportunities', *Technological Forecasting & Social Change*, 69, 625-639.

Leah Eisenberg, "Medicating Death Row Inmates So They Qualify for Execution," AMA Case in Health Law, Vol. 6, No. 9, September 2004.

Ledley, Fred D. 1995. "Nonviral Gene Therapy: The promise of Genes as Pharmaceutical Products," *Human Gene Therapy*, Vol. 6, Sep, pp 1129-1144.

Lee, M.S. and M. Head-Gordon. 2000. Absolute and relative energies from polarized atomic orbital selfconsistent field calculations and a second order convergence with size and composition of the secondary basis. *Computers and Chemistry*. Vol. 24(3,4). 295-301.

Lee, S.C. 1998. The nanobiological strategy for construction of nanodevices. *Biological molecules in nanotechnology: the convergence of biotechnology, polymer chemistry and materials science*, edited by S.C. Lee and L. Savage (IBC Press, Southborough, MA).

Liang, G., Z. Li, and E. Wang. 1996. *J. Mater. Sci.* Makino, A., A. Inoue, T. Hatanai, and T. Bitoh. 1997. *Materials Science Forum* 235-238: 723.

Lipson, H. and J. B. Pollack, Nature, 406, 974-978, (2000). *Automatic design and manufacture of robotic life forms.* See, also, several additional articles on this subject in contemporary issues of Nature..

M.L. Petrillo, C.J. Newton, R.P. Cunningham, R.-I. Ma, N.R. Kallenbach and N.C. Seeman. The Ligation and Flexibility of 4-Arm DNA Junctions. Biopolymers, 27, 1337-1352 (1988).

Mackenzie, D.: 1996a, 'Economic and Sociological Explanations of Technological Change', in: *Knowing Machines: Essays on Technical Change*, MIT Press, Cambridge, MA, pp. 49-65.

Mackenzie, D.: 1996b, 'Marx and the Machine', in: *Knowing Machines: Essays on Technical Change*, MIT Press, Cambridge, MA, pp. 23-47.

Malinow, R., Z.F. Mainen, and Y. Hayashi. 2000. LTP mechanisms: from silence to four-lane traffic. *Curr. Opin. Neurobiol.* 10(3): p. 352-7.

Mao, C., W. Sun, and N.C. Seeman. 1997. Construction of Borromean rings from DNA. *Nature* 386: 137-138.

Mark, H., Statement of The Honorable Hans Mark, Director, Defense Research and Engineering, before the House Armed Service Committee, October 20, 1999.

Martin, T.P., U. Naher, H. Schaber, U. Zimmerman. 1993. *Phys. Rev. Lett.* 70: 3079.

Marty, A.: 2003, *Testimony before the Committee on Science on HR 766, Nanotechnology Research and Development Act*, US Government Printing Office, Washington, DC.

Matsuki, H., K. Ofuji, N. Chubachi, and S. Nitta. 1996. Signal transmission for implantable medical devices using figure-of-eight coils. *IEEE Trans. on Magnetics.*, vol. 32. No.5, 5121. September.

McConnell, S.A., ed. 1998. *Pharmaceuticals.* Dun and Bradstreet Industry Reference Handbooks, Gale, Detroit, pp. 30-34.

McCulloch, W.S.: 1962, 'The imitation of one form of life by another – Biomimesis', in: E.E. Bernard & M.R. Kare (eds.), *Biological prototypes and synthetic systems*, Plenum Press, New York, vol. 1., p. 393-97.

Meadows, D.L., et al. 1972. *The Limits to Growth*, New York: University.

Mehl, R.F., and R.W. Cahn. 1983. Historical development. In *Physical metallurgy.* North Holland.

Milligan, W.W., S.A. Hackney, M. Ke, and E.C. Aifantis. 1993. *Nanostructured Materials* 2: 267.

Mehta, M.D.: 2004, 'From Biotechnology to Nanotechnology: What Can We Learn from Earlier Technologies?, *Bulletin of Science, Technology & Society*, 24, 34-39.

Merz, J.L. 1986. "The Optoelectronics Joint Research Laboratory: Light Shed on Cooperative Research in Japan." *Scientific Bulletin* (ONR Far East Office). 11 (4), 1-30 (Oct.-Dec.).

Meyer, G. and Amer, N.M. (1988) Novel optical approach to atomic force microscopy. Appl. Phys. Lett. 53(12), 1045-1047

Meyer, G. and Amer, N.M. (1990) Simultaneous measurement of lateral and normal forces with an optical-beam-deflection atomic force microscope. Appl. Phys. Lett. 57(20), 2089-2091

Meyer, M. & Kuusi, O.: 2004, 'Nanotechnology: Generalizations in an Interdisciplinary Field of Science and Technology', *Hyle: International Journal for Philosophy of Chemistry*, 10(2), 153-168.

Michael E.J. Holwill, Peter Satir, "II.4 Generation of Propulsive Forces by Cilia and Flagella," in J. Bereiter-Hahn, O.R. Anderson, W.-E. Reif, eds., *Cytomechanics: The Mechanical Basis of Cell Form and Structure*, Springer-Verlag, Berlin, 1987, pp. 120-130.

Michael Page, Donald W. Brenner, "Hydrogen abstraction from a diamond surface: *Ab initio* quantum chemical study using constrained isobutane as a model," *J. Am. Chem. Soc.* 113(1991): 3270-3274.

Mishra, R.S., R.Z. Valiev, and A.K. Mukherjee. 1997. *Nanostructured Materials* 9: 473. Morris, D.G., and M.A. Morris. 1991. *Acta Metall. Mater.* 39: 1763-1779.

Mizrach, S., *Should there be a limit placed on the integration of humans and computers and electronic technology?*

Mnyusiwalla, A.; Daar, A.S.; Singer, P.A.: 2003, '"Mind the Gap": Science and Ethics in Nanotechnology, *Nanotechnology*, 14, R9-R13.

Morris, D.G., and M.A. Morris. 1997. *Materials Science Forum* 235-238: 861. Nagpal, P., and I. Baker. 1990. *Scripta Metall. Mater.* 24: 2381.

Mui, C. and C. Musgrave, "The Hydroxylation and Oxidation of the Ge(100)-2x1 Surface by Hydrogen Peroxide and Water," *Langmuir*, 20, 7604-7609 (2004).

Perkal M., C. Marks, M.I. Lorber, W.H. Marks. 1992. "A three-year experience with serum anodal trypsinogen as a biochemical marker for rejection in pancreatic allografts. False positives, tissue biopsy, comparison with other markers, and diagnostic strategies." *Transplantation*, Feb; 53 (2): pp. 415-419.

Philip Ball, "Synthetic Biology: Starting from Scratch," Nature, 431, pp. 624-626, 7 October 2004.

Philip G. Collins, Michael S. Arnold, Phaedon Avouris, "Engineering carbon nanotubes and nanotube circuits using electrical breakdown," *Science* 292(27 April 2001): 706-709.

Philip G. Collins, Phaedon Avouris, "Nanotubes for electronics," *Scientific American* 283 (December 2000): 62-69.

Philipse, A.P. 1998. "Colloidal Dispersions," section 3.4, A. ten Wolde (ed.), *Nanotechnology: Towards a molecular construction kit.* STT Report #60, The Hague, 171-178.

Pierce, N.A. and M.B. Giles. 2000. Adjoint reconstruction of superconvergent functionals from PDE approximations. *SIAM Rev.* Vol. 42(2). 247-264.

Pinch, T.: 1993, "Testing – One, Two, Three... Testing!': Toward a Sociology of Testing', *Science, Technology, and Human Values*, 18, 25-41.

Pitt, J.C.: 2004, 'The Epistemology of the Very Small', in: D. Baird, A. Nordmann & J. Schummer (eds.), *Discovering the Nanoscale*, Amsterdam: IOS Press, pp. 157-163.

Ramanan, V.R. 1998. Nanocrystalline soft magnetic alloys for applications in electrical and electronic devices. In *R&D Status and Trends*, ed. Siegel et al.

Ramos, L.; Lubensky, T.C.; Dan, N.; Nelson, P.; Weitz, D.A. 1999. "Surfactant-mediated two-dimensional crystallization of colloidal crystals", *Science*, 286, 2325-2328.

Reed, Mark A. and James M. Tour. 2000. "Computing with Molecules." *Scientific American.* June.. SIA (Semiconductor Industry Association). 1999. "1999 International Technology Roadmap for Semiconductors."

Rheinberger, H.-J.: 1997, *Toward a History of Epistemic Things: Synthesizing Proteins in the Test Tube*, Stanford University Press, Stanford.

Rietman, E.A.: 2001, 'Drexler hypothesis of a universal assembler is supported not by theoretical arguments alone but by existence proof in the form of biological life', in: *Molecular Engineering of Nanosystems*, Springer, New York & Berlin.

Rinzler, A.G.; Liu, J.; Dai, H.; Nikolaev, P.; Huffman, C.B.; Rodriguez-Macias, F.J.; Boul, P.J.; Lu,

A.H.; Heymann, D.; Colbert, D.T.; Lee, R.S.; Fischer, J.E.; Rao, A.M.; Eklund, P.C.; Smalley, R.E. 1998.

Roco, M.&Bainbridge, W. (eds.). Converging Technologies for Improving Human Performances, National Science Foundation report, 2002.

Roco, M.C. & Tomellini, R. (eds.): 2002, *Nanotechnology: Revolutionary Opportunities and Societal Implications, Workshop, Lecce (Italy), 31 January—1 February 2002*, Luxemburg: Office for Official Publications of the European Communities, [200 pp.]

Roco, M.C., R.S. Williams, and P. Alivisatos (eds). 1999. *Nanotechnology Research Directions: IWGN Workshop Report, Vision for Nanotechnology R&D in the Next Decade*. (Kluwer Academic Publishers.

Roco, M.C.: 2003a, 'Broader Societal Issues of Nanotechnology', *Journal of Nanoparticle Research*, 5, 181-189.

Roco, M.C.: 2003b, 'Coherence and Divergence in Megatrends in Science and Engineering', in: M.C. Roco, W.S. Bainbridge (eds.), *Converging Technologies for Improving Human Performance*, Kluwer, Dordrecht, pp. 79-96.

Roco, M.C.: 2003c, 'Fundamentally New Manufacturing Processes and Products', in: Roco, M.C. & Bainbridge, W.S., (eds.), *Converging Technologies for Improving Human Performance*, Kluwer, Dordrecht, pp. 300-302.

Roco, M.C.: 2004, 'The Emergence and Policy Implications of Converging Technologies'

Roco, M.C.; Bainbridge, W.S. (eds.): 2002, *Converging Technologies for Improving Human Performance: Nanotechnology, Biotechnology, Information Technology and the Cognitive Science*, Arlington, VA: National Science Foundation.

Rodney Brooks, "The Merger of Flesh and Machines," The Next Fifty Years: Science in the First Half of the Twenty-First Century, ed. by John Brockman, 2002, p. 191.

Rohrer, H.: 1995, 'The Nanometer Age – Challenge and Chance', *Microelectronic Engineering*, 27, 3-15.

Rosenberg, Nathan. 1994. *Exploring the Black Box: Technology, Economics, and History*. New York: Cambridge University Press.

Rothman, B.K.: 1998, *Genetic Maps and Human Imagination*, W.W. Norton, New York & London,

Ruthven, D.M., S. Farooq, K.S. Knaebel. 1994. *Pressure swing adsorption*. New York: VCH Publishers.

Ruthven, D.M., S. Farooq, K.S. Knaebel. 1994. *Pressure swing adsorption*. New York: VCH Publishers. Sayari, A. 1996. *Chem. Mater*. 8: 1840.

Sankey, H.: 1994, *The Incommensurability Thesis*, Avebury, Aldershot.

Santini, John T., Jr., Michael J. Cima and Robert Langer. 1999. "A Controlled-release Microchip" *Nature*, Vol 397, 28 Jan, p 335-338.

Sarewitz, D.; Woodhouse, E.: 2003, 'Small is Powerful', in: A. Lightman, D. Sarewitz & Chr. Desser, (eds.), *Living with the Genie: Essays on Technology and the Quest for Human Mastery*, Washington, DC: Island Press, pp. 63-83.

Sarikaya, M.; Aksay, I. (eds.), 1995, *Biomimetics. Design and Processing of Materials*, AIP Press, Woodbury, New York.

Schmidt, J.C.: 2004, 'Unbounded Technologies: Working Through Technological Reductionism of Nanotechnology', in: D. Baird, A. Nordmann & J. Schummer (eds.), *Discovering the Nanoscale*, Amsterdam: IOS Press, pp. 35-50.

Schuler, E.: 2004, 'Perception of Risks and Nanotechnology', in: D. Baird, A. Nordmann & J. Schummer (eds.), *Discovering the Nanoscale*, Amsterdam: IOS Press, pp. 279-284.

Schummer, J.: 2001, 'Ethics of Chemical Synthesis', *Hyle: International Journal for Philosophy of Chemistry*, 7, 103-124 [online: www.hyle.org/journal/issues/7/schummer.htm].

Schummer, J.: 2003, 'Aesthetics of Chemical Products: Materials, Molecules, and Molecular Models', *Hyle: International Journal for Philosophy of Chemistry*, 9, 73-104 [online: www.hyle.org/journal/issues/9-1/schummer.htm].

Schummer, J.: 2003, 'The Notion of Nature in Chemistry', *Studies in History and Philosophy of Science*, 34 (2003), 705-736.

Schummer, J.: 2004, 'Interdisciplinary Issues in Nanoscale Research', in: D. Baird, A. Nordmann, J. Schummer (eds.), *Discovering the Nanoscale*, IOS Press, Amsterdam, pp. 9-20.

Schwarz, A.E.: 2004, 'Shrinking the 'Ecological Footprint' with NanoTechnoScience?', in: D. Baird, A. Nordmann & J. Schummer (eds.), *Discovering the Nanoscale*, Amsterdam: IOS Press, pp. 203-208.

Schwarz, R.B. 1998. Storage of hydrogen in powders with nanosized crystalline domains. In *R&D status and trends*, ed. Siegel et al.

Siegel et al. Ogunnaike, B and W. Ray. 1994. *Process dynamics, modeling and control*. Oxford University Press, pp 5-21; 1033-48.

Siegel R.W. and G.E. Fougere. 1994. In *Nanophase materials*, ed. G.C. Hadjipanayis and R.W. Siegel. Netherlands: Kluwer Acad. Publ.

Siegel, H.: 1980, 'Objectivity, Rationality, Incommensurability, and More', *British Journal for the Philosophy of Science*, 31, 359-384.

Siegel, R.W., E. Hu, and M.C. Roco, eds. 1998. *R&D status and trends in nanoparticles, nanostructured materials, and nanodevices in the United States*. Baltimore: Loyola College, International Technology Research Institute. NTIS #PB98-117914.

Siegel, R.W., E. Hu, and M.C. Roco, eds. 1998. *R&D status and trends in nanoparticles, nanostructured materials, and nanodevices in the United States*. Baltimore: Loyola College, International Technology Research Institute.

Smalley, R.E. 2000. Hearing of the Senate Science and Technology Caucus on Nanotechnology. April 5, (unpublished).

Smalley, R.E.: 1999, 'Prepared written statement and supplemental material', Rice University, 22 June.

Smalley, R.E.: 2001, 'Of Chemistry, Love and Nanobots', *Scientific American*, (Sept.), 76-77.

Smith, A., *The Wealth of Nations: An Inquiry into the Nature and Causes*, Random House, New York, 1994.

Smith, D., H. Babcock, and S. Chu. 1999. Single-polymer dynamics in steady shear flow. *Science*. Vol. 283. 1724-1727. March.

Snider, G.L.; Orlov, A.O.; Amlani, I.; Zuo, X.; Bernstein, G.H.; Lent, C.S.; Merz, J.L.; Porod, W. 1999. "Quantum-dot cellular automata: Review and recent experiments (invited)", *Journal of Applied Physics*, 85, 4283-4285.

Snow, E.S., P.M. Campbell, and F.K. Perkins. 1997. Nanofabrication with proximal probes. *Proceedings of the IEEE* 85: 601-11.

Sterman, J. D. (1989b). Modeling Managerial Behavior: Misperceptions of Feedback in a Dynamic Decision Making Experiment. *Management Science*, 35(3), 321-339.

Steve, Jurvetson, "Transcending Moore's Law with Molecular Electronics," Nanotechnology Law & Business Journal, Vol. 1, No. 1, article 9, p. 9.

Stipe, B.C.; Rezaei, M.A.; Ho, W. 1998. "Single-molecule vibrational spectroscopy and microscopy", *Science*, 280, 1732-1735.

Stix, G.: 2001, 'Little Big Science: Nanotechnology Is All the Rage. But Will It Meet Its Ambitious Goals? And What The Heck Is It?', *Scientific American* 285, 32-37.

Stokes, Donald E. 1997. *Pasteur's Quadrant: Basic Science and Technological Innovation*. Washington, D.C.: Brookings Institution Press.

Stranick, S.J.; Atre, S.V.; Parikh, A.N.; Wood, M.C.; Allara, D.L.; Winograd, N.; Weiss, P.S. 1996. "Nanometer-scale phase separation in mixed composition self-assembled monolayers", *Nanotechnology*, 7, 438-442.

Tenner, E.: 2001, 'Nanotechnology and Unintended Consequences', in: M.C. Roco and W.S. Bainbridge (eds.), *Societal Implications of Nanoscience and Nanotechnology*, Dordrecht: Kluwer, pp. 241-246.

Tighe, T.S., J.M. Worlock, and M.L. Roukes. 1997. Direct thermal conductance measurements on suspended monocrystalline nanostructures. *Appl. Phys. Lett.* 70: 2687-9.

Tihamer Toth-Fejel, "Modeling kinematic cellular automata: An approach to self-replication," Environmental Research Institute of Michigan, Ann Arbor, MI, 4 October 2000.

Tilton, John E. (ed). 1991. *World Metal Demand*, Washington, DC: Resources for the Future.

Tobio, M., R. Gref, A. Sanchez, R. Langer and M.J. Alonso. 1998. "Stealth PLA-PEG Nanoparticles as Protein Carriers for Nasal Administration", *Pharmaceutical Research*, Vol. 15, No 2, pp 270-275.

Tramontin, A.D. and E.A. Brenowitz. 2000. Seasonal plasticity in the adult brain. *Trends Neurosci*. 23(6): p. 251-8.

Trivelli, A., and W.F. Smith. 1939. *Photog. J.* 79: 330,463,609. As referred to in T.H. James. 1977. *The theory of the Photographic process.* New York: MacMillan (p. 100).

4

Advances in Nano Research

Societal Approaches for Assessing Nanotechnology's Implications

Nanotechnology is seen by some investors as nirvana, by others as a fool's paradise. The truth is both more complex and more prosaic. The field has been hyped, potentially to its own detriment. But for those who can separate the science from the fiction, it's a promising arena for long and short-term investments, which is already cutting costs and improving products and processes in countless sectors, enhancing everything from cosmetic creams to fuel efficiency. Nanotechnology's vast reach helps explain why it's so difficult to nail down its actual potential. The market is ill-defined, misunderstood and easily distorted. "Nanotechnology is a huge building and there are so many doors," says Patrick Hunziker, a cardiologist working on medical applications of nanotechnology at the University of Basel hospital in Switzerland. "Every one you open leads to so many more questions." Commercial success lingers behind some of those doors, but the impulse to add the 'nano' suffix to a business plan only to boost its appeal can lead investors and entrepreneurs astray.

This study will explore the secrets of setting up a successful nanotechnology company, and reveal some of nanotechnology's unique hurdles and winning strategies. The report breaks down into four sections. The first defines our terms, and examines what nanotechnology is. The second examines the commercial potential of nanotechnology. The third identifies the reasons why nanotechnology is so complex. And the fourth and final section provides practical advice on how start-ups can win in this space.

"There are a few areas that from time to time demand commentary from a venture capitalist and nanotechnology is one of them. We have drawn on the knowledge and expertise from our considerable international network to shed light on what it takes to create a successful business in the field of nanotechnology. Furthermore, through our relationship with the Institute of Nanotechnology, we have brought together an incredible list of international contributors to this paper. In summary, we believe this paper provides a unique and commercial insight into this exciting field".

"The Institute of Nanotechnology is recognised as the focus for activity on the nanoscale, having been involved since 1994 in bringing together the nanotechnology community worldwide. Using the contacts of the Institute, 3i has brought its experience and knowledge to bear in this very fragmented and unfortunately much hyped field, to identify where real business potential lies and the realities of what is on offer".

Molecular Magic

Novel things happen when moving from the micro level (one micrometre is one millionth of a metre) to the nano level. By working at the nanoscale, materials can exhibit new properties from those exhibited by bulk material. Working with particles that are less than 100 nanometres in size profoundly changes the particle's percentage of exposed area in proportion to its total volume when compared with bulk materials. "In particles of three to five nanometres in size, about a third of the atoms are surface atoms," explains Peter Dobson, a professor of Engineering Science at Oxford University, and the founder of several nanotechnology start-ups. By comparison, in an element with the diameter of a human hair, only a tiny fraction of a percentage of the element's atoms are located on its surface. "Nanoparticles in the three-to-five nanometre range behave a lot like gas particles" rather than solids, adds Professor Dobson, floating where their bigger counterparts would sink. Nanoparticles also dissolve more readily than the same materials in bulk. This extra exposed surface area changes how elements and nanoparticles interact with one another. Consider how useful nano-sized titanium dioxide would be for suncream, for instance, since it absorbs more ultra-violet light than its bulk counterpart. Sunblock with titanium-dioxide particles at sizes below 50 nanometres in diameter also lasts longer on the skin, explains Professor Dobson. Oxonica, a UK spin-out from Oxford University that Professor Dobson co-founded, is one of several firms working on sunblock that contains such nanoparticles. Being able to control such attributes enables nanotechnology to deliver improved functionality. Biological membranes with nano-sized pores, for example, let needed fluids flow freely but block out oversized antibodies or damaging digestive enzymes. In another example of size making a difference, C Sixty, a Toronto-based start-up which is moving its headquarters to Houston, is working on several drugs delivered by hollow buckyballs – or Buckminsterfullerenes, a carbon form discovered in 1985 – just one nanometre large. The tiny carriers are small enough for the kidneys to excrete so that they pass safely from the body when their work is done.

When trying to pinpoint the start of nanotechnology, many scientists look to Richard Feynman as a founding father. His famous "there's plenty of room at the bottom" speech in 1959 foretold many of today's developments. But Norio Taniguchi, a professor from Tokyo Science University, was the first to coin the term 'nanotechnology,' although many mistakenly give the credit to Eric Drexler's Engines of Creation of a dozen years later. (According to a paper written by four members of the European Society for Precision Engineering and Nanotechnology – and published in the annals of the International Institution for Production Engineering Research – Professor Taniguchi used the term at an engineering conference in 1974 to describe ultra-fine machining.)

But the practice of nanoscience only began to take off once scientists could view nanoparticles, starting with the development of the scanning tunnelling microscope (STM) at IBM's Zurich Research Laboratory in the early 1980s. In an STM, a tiny stylus – whose tip is only one-atom large – passes just above the surface of a metal sample to produce an atomic-scale, three-dimensional representation of the surface. By the time IBM scientists Gerd Binnig and Heinrich Rohrer won the 1986 Nobel prize for physics for their STM design, Professor Binnig was already back at the coalface: along with Christoph Gerber and Calvin Quate, he devised the atomic force microscope (AFM) for working with non-conductive materials that same year. An AFM generates images based on the movement in the microscope's tip, which is attached to a flexible cantilever and is brought within nanometres of a non-conductive element's surface. Today, STMs and AFMs let scientists not only observe but also manipulate matter on a nanoscale. "They are fantastic tools that have enabled nanotechnology to develop," says Cambridge University's Professor Welland. With the STM and the AFM "we've moved from looking to fabrication and exploitation."

Alarm bells are starting to ring throughout the whole nanotechnology field that undoubted promise is shading into unjustified hype. Investors need to be crystal-clear about start-ups' commercial potential. The biggest roadblock to progress in nanotechnology today, according to 3i survey respondents, is the challenge of finding and recognising commercial applications of research. "There is huge potential out there, but the emphasis is still on science," says Dr Gerber, a physicist at IBM's Zurich Research Laboratory, and co-inventor of the atomic force microscope. Science explores concepts, rules and behaviours, while technology applies what science has learned. "Everyone talks about nanotechnology," says Professor Emeritus Pat McKeown, a founding member of the European Society for Precision Engineering and Nanotechnology. "But most people in the field are still working on nanoscience. Nanoscience is nanotechnology in its infancy."

Hence the 'handle with care' warning that should be attached to the various estimates of nanotechnology market size. In a 2001 report, for instance, the National Science Foundation presents statistics from a variety of sources indicating that by 2011, nanostructured materials and processes will be worth $340bn annually and nanotechnology-enabled aerospace will be worth $70bn, while nanotechnology will affect more than $180bn-worth of business in the pharmaceuticals sector within 10 to 15 years. Such estimates do help demonstrate the scale of the potential opportunity. But there's a lot of speculation behind them, triggering alarm bells throughout the entire nanotechnology industry that promise is shading into hype. In the 3i survey, the market force considered most likely to cause a backlash against nanotechnology is the notion that it can fix everything. Professor Scott Manalis of the Massachusetts

Institute of Technology (MIT) says there is "absolutely" a danger to nanotechnology's advancement if its proponents over-promote its promise. "Nanotechnology has an increasingly large number of definitions," notes Professor Manalis. "In some cases, the public interpretation is quite different than that of the actual research field." This disjuncture between reality and public expectations often comes about because the most spectacular examples of nanotechnology at work, many of them in the healthcare sector, are also often the furthest from commercialisation. "People hype developments with the biggest impact, but those are the developments that are really far away," says Mike Warren, a consultant with BPA Consulting, a firm of UK electronic and optoelectronic technology analysts. "You can easily get confused by people jumping on the nanotechnology bandwagon," agrees Professor Atherton, former managing director of UK nanotechnology firm Queensgate Instruments and now an angel investor.

Fortunately, the connection many observers now make between nanotechnology and the dotcom era is a flawed one. Dotcom euphoria was sustained principally by hype in the consumer market. Nanotechnology ventures are focused on business-to-business (B2B) applications for often specialised clients. Hype or no hype, nanotechnology ventures with nothing new to offer their customers will find it very difficult to flourish. As a result, nanotechnology start-ups need to be crystal-clear about the commercial benefits they can offer. Those benefits tend to coalesce into one of two propositions: new products and procedures with innovative functionality, or improvements to existing products and procedures through faster and/or cheaper processes. Take the carbon nanotube, a form of carbon that NEC's Sumio Iijima discovered in 1991.

One thousand times longer than wide, a carbon nanotube looks a bit like chicken wire rolled into a tube and has a multitude of uses. Single-wall carbon nanotubes can conduct electricity 100,000 times greater than graphite and carbon fibres. Single-wall carbon nanotubes are also 100 times stronger than steel and are now being added to many materials such as plastic. One gram or less of single-walled carbon nanotubes can light up a 20 inch computer screen, says Bob Gower, chief executive officer of Houston-based Carbon Nanotechnologies (CNI), which produces about 25 grams of single-walled carbon nanotubes daily and sells them at between $500 and $750 per gram. The material is also used to strengthen the Kevlar used in bulletproof vests and to shield battlefield communication systems from interference or snooping, he says. Mitsui is clearly convinced of the commercial potential of nanotubes. The Japanese company is building a multi-wall carbon nanotube factory in Tokyo. Built in co-operation with Mitsui's subsidiary, the Carbon Nanotech Research Institute, the plant will begin production by October 2002. Starting with an annual production level of 120 tons, the plant will produce multi-wall carbon nanotubes for customers such as the makers of electronic devices, automobiles and paint.

Despite the nanotechnology naysayers, it is indisputable that activity in this space is accelerating fast. Consider the following:

- Government spending on nanotechnology projects in Japan more than quadrupled over the last four years, jumping to $466m in 2001 from $113m in 1998, according to Naoya Kobayashi at the Hitachi Research Institute in Tokyo.
- The US government launched its National Nanotechnology Initiative (NNI) in fiscal year 2000 with a budget of $270m and then boosted the budget by 56% to $422m in 2001. The numbers keep going up. In 2002, $604m was spent on nanotechnology research and development, according to the NNI, and a budget request of over $710m has been submitted for 2003. Says Mr Warren of BPA: "When the US government is willing to spend $710m a year on nanotech research, it's a sign that a lot of people are becoming very serious about the possibilities of the technology."
- European investment is more modest although close comparisons between EU and US spending can be misleading given that nanotechnology is not always defined in the same way. The important thing to note is that the trend is the same – upwards. Government support for nanotechnology from the European Commission and EU member states has increased by more than 40% over three years, from about 130m in 1997 to 184m in 2000, according to figures from Ramon Compañó, a technology co-ordinator at the European Commission. He estimates that government support in 2001 rose even further, to 300m.

Like governments, corporations are also investing heavily. Mitsui of Japan is investing $77m in nanotechnology over five years while Hitachi is investing between $38.5m and $53.8m annually, according to Mr Kobayashi. Many companies have also launched private equity funds to invest in nanotechnology start-ups. Mitsubishi's Nanotech Partners fund, for example, has ¥15bn ($120m) to invest in nanotech ventures between now and 2005.

Patent information from the Derwent World Patents Index confirms this pattern of accelerating activity. The number of patents being published per year from the 40 issuing authorities covered by Derwent shows explosive growth from 1998 onwards. Growth is consistent across the years, with North America, for example, earning four times the number of patents in 2001 as it did in 1997.

Cost savings are one reason why nanotechnology is attracting increasing investment. General Motors is using lighter, stronger materials based on nanoparticles in some of its mid-sized vans, thanks to partnerships with Basell, Southern Clay Products and Blackhawk Automotive Plastics. GM has its own nanotechnology research and development team of five scientists in-house, but with help from its partners has achieved a 7-21% weight saving on certain car parts by adding a thermoplastic olefin nanocomposite to plastic. "It's cheaper and lighter than using steel, which is of course the holy grail for the automotive industry," says Tim Harper, chief executive at CMP Científica, a research firm based in Madrid. "If you have a weight saving, you have a fuel saving." In another field entirely, current DNA testing for disease detection and prevention is expensive and requires a lot of bulky machinery. Professor Manalis at MIT is working on a device that can detect DNA sequences in very small volumes. A tiny transistor is located on the end of an AFM cantilever, explains Professor Manalis, and the device senses molecular charge at the end of the cantilever. Nanotechnology is already driving down the cost and boosting the convenience of medical testing, doing away with vials of blood shipped off to a lab for slow testing on expensive equipment. Now, in-office tests use pinprick quantities of blood for instant results. "Nanotechnology allows an array of diagnosis with a minimum of material," says Dr Hunziker, the University of Basel cardiologist. "The material cost becomes very low and you get bedside results within minutes."

Clearly there is a risk that hype could over-inflate expectations about nanotechnology. But there is a competing risk that hype could scare investors away from genuine commercial opportunities. Fortunately, that doesn't seem to be happening. Governments, corporates and investors are all contributing to a pattern of accelerating activity in the field. With an estimated 25 nanotechnology start-ups already listed in the US and UK, and large manufacturing initiatives like Mitsui's well under way, it's hard to disagree with the conclusion of Hans Coufal, director of science and technology at IBM's Almaden Research Center in California: "Nanotechnology is not the future – it is here today".

Mapping the Market

New firms have to find a market focus among the multiple disciplines, locations, and timeframes the nanotechnology industry encompasses. Every new business faces challenges. But nanotechnology is a particularly complex field. "If you want to make a business out of nanotechnology, you've got some unique challenges," says Kevin O'Grady, who founded magnetic-fluids supplier Liquids Research in the UK ten years ago. None is more challenging than understanding a field that embraces so many disciplines, locations, and timeframes. Finding the appropriate market focus – or to put it another way, being in the right place with the right people at the right time – is a precondition of success.

The right people Nanotechnology-which many see as the crossroads of chemistry, physics, biology and material science-requires scientists from multiple disciplines to collaborate. "It's difficult because people have to learn how to communicate with one another," says Jean-Charles Guibert, programme director at France's Minatec, a nanotechnology centre being developed near Grenoble. "You cannot be successful [in nanotechnology] with chemistry or engineering alone," adds Ralf Zastrau, chief executive officer of Germany's Nanogate Technologies. "You have to have all the disciplines working together." Universities have therefore started to offer cross-disciplinary training programmes. Beginning in autumn 2002, for example, the University of Basel in Switzerland will let students work towards a bachelor's degree in nanotechnology, while the University of Cambridge is considering offering a masters in the field.

Nanotechnology research centres, which seem to be popping up everywhere, are another potential source of talent. In the US, for example, research centres are being built at Columbia, Cornell, Harvard, Northwestern and Rice Universities, among others. In Europe, the Micro and Nanotechnology Innovation Centre (Minatec) – which may well turn out to be Europe's largest nanotechnology centre when it is finished in 2005 – is being built to support some 3,000 researchers, students and business advisers. Mixing the correct blend of scientific expertise is only half the story, of course. As in other fields, successful nanotechnology companies must also call on astute commercial heads to run the business side of the operation efficiently, shepherding budgets through multi-year product development cycles and turning technological concepts into revenue-generating products. *The right place* Location is clearly important in giving nanotechnology companies easy access to a pool of qualified people. But it also helps determine access to capital in the form of investors, and access to market in the form of customers and partners. Certain key regions and countries dominate activity. According to the Derwent World Patents Index, North America was awarded 2,716 nanotechnology-related patents between 1995 and 2001, while Asia earned 2,213 during the same time period. Europe lagged a little with 1,873 published patents. Within the Asian region, Japan is the leading nation for patent output ahead of China and Korea. Within Europe, Germany has the lead over France and the United Kingdom. Geographical specialisations are also emerging within the field. Our survey respondents were asked to identify the countries in which the most sophisticated nanotechnology developments in particular industries are happening. Unsurprisingly, the United States came out top for every industry, but by splitting the US into regions, a more complex picture emerges. Survey respondents consider Japan to be the global leader in electronic applications of nanoscience, Germany to be the pacesetter in chemical applications and the US west coast to lead in medical, material and manufacturing applications of nanoscience.

The right time A lot of the money flowing to nanotechnology start-ups and research centres now won't be able to support firms long enough for them to achieve their commercial goals. Areas such as pharmaceuticals and the quest to replace silicon with organic materials in semiconductors will remain in R&D for years to come. For example, the average drug can take 16 years to journey from discovery to market delivery, says Dr Uri Sagman, president and chief executive officer of C Sixty. Targeting the US's $4bn market for AIDS drugs and the $30bn market for cancer drugs, C Sixty will be lucky if – with fast track approval in the US – it can get its drugs on the market four years from now. By contrast, there are areas of early adoption where nanotechnology firms are already selling products. The materials and composites market is one such area-in our survey, smart paints, pigments and coatings were considered the fastest adopter of nanotechnology and the most promising commercial area over the next five years. Take Germany's Nanogate Technologies, for instance, which already has over a dozen such products on the market. "Engineered surfaces have tremendous benefits and no regulatory barriers" when compared with fields such as medicine, notes Professor Atherton, the angel investor. Cosmetics is a pace-setter for just this reason: French giant L'Oreal is one of the world's top holders of nanotechnology-based patents, according to Nick Fox-Male of Eric Potter Clarkson, a UK intellectual-property specialist. These time-to-market variations are not hard and fast, of course. Investors have already exited successfully from drug delivery companies, for example. What's more, even where timescales are longer, there's opportunity for suppliers to the nanotechnology industry to exploit. Poor availability of instruments to manufacture and analyse nanomaterials feature in the list of industry roadblocks identified by survey respondents. As a result, enabling products—the tools and techniques that facilitate nanotechnology development and manufacturing—can expect immediate demand.

To be successful in nanotechnology, a number of ingredients have to be in perfect alignment. The team must combine different types of scientific expertise and a solid dose of business acumen. The location should enable access to staff, partners, investors and customers. And the time-to-market must be tailored to the resources available. If these elements are in place, a necessary, if not sufficient, condition of nanotechnology success will be too.

Model Nanotechnology Research and Development Act

Begun and held at the City of Washington on Tuesday, the seventh day of January, two thousand and three

To authorize appropriations for nanoscience, nanoengineering, and nanotechnology research, and for other purposes.

Be it enacted by the Senate and House of Representatives of the United States of America in Congress assembled,

This Act may be cited as the '21st Century Nanotechnology Research and Development Act'.

(a) NATIONAL NANOTECHNOLOGY PROGRAM- The President shall implement a National Nanotechnology Program. Through appropriate agencies, councils, and the National Nanotechnology Coordination Office established in section 3, the Program shall—

(1) establish the goals, priorities, and metrics for evaluation for Federal nanotechnology research, development, and other activities;

(2) invest in Federal research and development programs in nanotechnology and related sciences to achieve those goals; and

(3) provide for interagency coordination of Federal nanotechnology research, development, and other activities undertaken pursuant to the Program.

(b) PROGRAM ACTIVITIES- The activities of the Program shall include—

(1) developing a fundamental understanding of matter that enables control and manipulation at the nanoscale;

(2) providing grants to individual investigators and interdisciplinary teams of investigators;

(3) establishing a network of advanced technology user facilities and centers;

(4) establishing, on a merit-reviewed and competitive basis, interdisciplinary nanotechnology research centers, which shall—

(A) interact and collaborate to foster the exchange of technical information and best practices;

(B) involve academic institutions or national laboratories and other partners, which may include States and industry;

(C) make use of existing expertise in nanotechnology in their regions and nationally;

(D) make use of ongoing research and development at the micrometer scale to support their work in nanotechnology; and

(E) to the greatest extent possible, be established in geographically diverse locations, encourage the participation of Historically Black Colleges and Universities that are part B institutions as defined in section 322(2) of the Higher Education Act of 1965 (20 U.S.C. 1061(2)) and minority institutions (as defined in section 365(3) of that Act (20 U.S.C. 1067k(3))), and include institutions located in States participating in the Experimental Program to Stimulate Competitive Research (EPSCoR);

(5) ensuring United States global leadership in the development and application of nanotechnology;

(6) advancing the United States productivity and industrial competitiveness through stable, consistent, and coordinated investments in long-term scientific and engineering research in nanotechnology;

(7) accelerating the deployment and application of nanotechnology research and development in the private sector, including startup companies;

(8) encouraging interdisciplinary research, and ensuring that processes for solicitation and evaluation of proposals under the Program encourage interdisciplinary projects and collaborations;

(9) providing effective education and training for researchers and professionals skilled in the interdisciplinary perspectives necessary for nanotechnology so that a true interdisciplinary research culture for nanoscale science, engineering, and technology can emerge;

(10) ensuring that ethical, legal, environmental, and other appropriate societal concerns, including the potential use of nanotechnology in enhancing human intelligence and in developing artificial intelligence which exceeds human capacity, are considered during the development of nanotechnology by—

(A) establishing a research program to identify ethical, legal, environmental, and other appropriate societal concerns related to nanotechnology, and ensuring that the results of such research are widely disseminated;

(B) requiring that interdisciplinary nanotechnology research centers established under paragraph (4) include activities that address societal, ethical, and environmental concerns;

(C) insofar as possible, integrating research on societal, ethical, and environmental concerns with nanotechnology research and development, and ensuring that advances in nanotechnology bring about improvements in quality of life for all Americans; and

(D) providing, through the National Nanotechnology Coordination Office established in section 3, for public input and outreach to be integrated into the Program by the convening of regular and ongoing public discussions, through mechanisms such as citizens' panels, consensus conferences, and educational events, as appropriate; and

(11) encouraging research on nanotechnology advances that utilize existing processes and technologies.

(c) PROGRAM MANAGEMENT- The National Science and Technology Council shall oversee the planning, management, and coordination of the Program. The Council, itself or through an appropriate subgroup it designates or establishes, shall—

(1) establish goals and priorities for the Program, based on national needs for a set of broad applications of nanotechnology;

(2) establish program component areas, with specific priorities and technical goals, that reflect the goals and priorities established for the Program;

(3) oversee interagency coordination of the Program, including with the activities of the Defense Nanotechnology Research and Development Program established under section 246 of the Bob Stump National Defense Authorization Act for Fiscal Year 2003 (Public Law 107-314) and the National Institutes of Health;

(4) develop, within 12 months after the date of enactment of this Act, and update

every 3 years thereafter, a strategic plan to guide the activities described under subsection (b), meet the goals, priorities, and anticipated outcomes of the participating agencies, and describe—

(A) how the Program will move results out of the laboratory and into application for the benefit of society;

(B) the Program's support for long-term funding for interdisciplinary research and development in nanotechnology; and

(C) the allocation of funding for interagency nanotechnology projects;

(5) propose a coordinated interagency budget for the Program to the Office of Management and Budget to ensure the maintenance of a balanced nanotechnology research portfolio and an appropriate level of research effort;

(6) exchange information with academic, industry, State and local government (including State and regional nanotechnology programs), and other appropriate groups conducting research on and using nanotechnology;

(7) develop a plan to utilize Federal programs, such as the Small Business Innovation Research Program and the Small Business Technology Transfer Research Program, in support of the activity stated in subsection (b)(7);

(8) identify research areas that are not being adequately addressed by the agencies' current research programs and address such research areas;

(9) encourage progress on Program activities through the utilization of existing manufacturing facilities and industrial infrastructures such as, but not limited to, the employment of underutilized manufacturing facilities in areas of high unemployment as production engineering and research testbeds; and

(10) in carrying out its responsibilities under paragraphs (1) through (9), take into consideration the recommendations of the Advisory Panel, suggestions or recommendations developed pursuant to subsection (b)(10)(D), and the views of academic, State, industry, and other appropriate groups conducting research on and using nanotechnology.

(d) ANNUAL REPORT- The Council shall prepare an annual report, to be submitted to the Senate Committee on Commerce, Science, and Transportation and the House of Representatives Committee on Science, and other appropriate committees, at the time of the President's budget request to Congress, that includes—

(1) the Program budget, for the current fiscal year, for each agency that participates in the Program, including a breakout of spending for the development and acquisition of research facilities and instrumentation, for each program component area, and for all activities pursuant to subsection (b)(10);

(2) the proposed Program budget for the next fiscal year, for each agency that participates in the Program, including a breakout of spending for the development and acquisition of research facilities and instrumentation, for each program component area, and for all activities pursuant to subsection (b)(10);

(3) an analysis of the progress made toward achieving the goals and priorities established for the Program;

(4) an analysis of the extent to which the Program has incorporated the recommendations of the Advisory Panel; and

(5) an assessment of how Federal agencies are implementing the plan described in subsection (c)(7), and a description of the amount of Small Business Innovative Research and Small Business Technology Transfer Research funds supporting the plan.

(a) IN GENERAL- The President shall establish a National Nanotechnology Coordination Office, with a Director and full-time staff, which shall—

(1) provide technical and administrative support to the Council and the Advisory Panel;

(2) serve as the point of contact on Federal nanotechnology activities for government organizations, academia, industry, professional societies, State nanotechnology programs, interested citizen groups, and others to exchange technical and programmatic information;

(3) conduct public outreach, including dissemination of findings and recommendations of the Advisory Panel, as appropriate; and

(4) promote access to and early application of the technologies, innovations, and expertise derived from Program activities to agency missions and systems across the Federal Government, and to United States industry, including startup companies.

(b) FUNDING- The National Nanotechnology Coordination Office shall be funded through interagency funding in accordance with section 631 of Public Law 108-7.

(c) REPORT- Within 90 days after the date of enactment of this Act, the Director of the Office of Science and Technology Policy shall report to the Senate Committee on Commerce, Science, and Transportation, and the House of Representatives Committee on Science on the funding of the National Nanotechnology Coordination Office. The report shall include—

(1) the amount of funding required to adequately fund the Office;

(2) the adequacy of existing mechanisms to fund this Office; and

(3) the actions taken by the Director to ensure stable funding of this Office.

(a) IN GENERAL- The President shall establish or designate a National Nanotechnology Advisory Panel.

(b) QUALIFICATIONS- The Advisory Panel established or designated by the President under subsection (a) shall consist primarily of members from academic institutions and industry. Members of the Advisory Panel shall be qualified to provide advice and information on nanotechnology research, development, demonstrations, education, technology transfer, commercial application, or societal and ethical concerns. In selecting or designating an Advisory Panel, the President may also seek and give consideration to recommendations from the Congress, industry, the scientific community (including the National Academy of Sciences, scientific professional societies, and academia), the defense community, State and local governments, regional nanotechnology programs, and other appropriate organizations.

(c) DUTIES- The Advisory Panel shall advise the President and the Council on matters relating to the Program, including assessing—

(1) trends and developments in nanotechnology science and engineering;

(2) progress made in implementing the Program;

(3) the need to revise the Program;

(4) the balance among the components of the Program, including funding levels for the program component areas;

(5) whether the program component areas, priorities, and technical goals developed by the Council are helping to maintain United States leadership in nanotechnology;

(6) the management, coordination, implementation, and activities of the Program; and

(7) whether societal, ethical, legal, environmental, and workforce concerns are adequately addressed by the Program.

(d) REPORTS- The Advisory Panel shall report, not less frequently than once every 2 fiscal years, to the President on its assessments under subsection (c) and its recommendations for ways to improve the Program. The first report under this subsection shall be submitted within 1 year after the date of enactment of this Act. The Director of the Office of Science and Technology Policy shall transmit a copy of each report under this subsection to the Senate Committee on Commerce, Science, and Technology, the House of Representatives Committee on Science, and other appropriate committees of the Congress.

(e) TRAVEL EXPENSES OF NON-FEDERAL MEMBERS- Non-Federal members of the Advisory Panel, while attending meetings of the Advisory Panel or while otherwise serving at the request of the head of the Advisory Panel away from their homes or regular places of business, may be allowed travel expenses, including per diem in lieu of subsistence, as authorized by section 5703 of title 5, United States Code, for individuals in the government serving without pay. Nothing in this subsection shall be construed to prohibit members of the Advisory Panel who are officers or employees of the United States from being allowed travel expenses, including per diem in lieu of subsistence, in accordance with existing law.

(f) EXEMPTION FROM SUNSET- Section 14 of the Federal Advisory Committee Act shall not apply to the Advisory Panel.

(a) IN GENERAL- The Director of the National Nanotechnology Coordination Office shall enter into an arrangement with the National Research Council of the National Academy of Sciences to conduct a triennial evaluation of the Program, including—

(1) an evaluation of the technical accomplishments of the Program, including a review of whether the Program has achieved the goals under the metrics established by the Council;

(2) a review of the Program's management and coordination across agencies and disciplines;

(3) a review of the funding levels at each agency for the Program's activities and the ability of each agency to achieve the Program's stated goals with that funding;

(4) an evaluation of the Program's success in transferring technology to the private sector;

(5) an evaluation of whether the Program has been successful in fostering interdisciplinary research and development;

(6) an evaluation of the extent to which the Program has adequately considered ethical, legal, environmental, and other appropriate societal concerns;

(7) recommendations for new or revised Program goals;

(8) recommendations for new research areas, partnerships, coordination and management mechanisms, or programs to be established to achieve the Program's stated goals;

(9) recommendations on policy, program, and budget changes with respect to nanotechnology research and development activities;

(10) recommendations for improved metrics to evaluate the success of the Program in accomplishing its stated goals;

(11) a review of the performance of the National Nanotechnology Coordination Office and its efforts to promote access to and early application of the technologies, innovations, and expertise derived from Program activities to agency missions and systems across the Federal Government and to United States industry;

(12) an analysis of the relative position of the United States compared to other nations with respect to nanotechnology research and development, including the identification of any critical research areas where the United States should be the world leader to best achieve the goals of the Program; and

(13) an analysis of the current impact of nanotechnology on the United States economy and recommendations for increasing its future impact.

(b) STUDY ON MOLECULAR SELF-ASSEMBLY- As part of the first triennial review conducted in accordance with subsection (a), the National Research Council shall conduct a one-time study to determine the technical feasibility of molecular self-assembly for the manufacture of materials and devices at the molecular scale.

(c) STUDY ON THE RESPONSIBLE DEVELOPMENT OF NANOTECHNOLOGY- As part of the first triennial review conducted in accordance with subsection (a), the National Research Council shall conduct a one-time study to assess the need for standards, guidelines, or strategies for ensuring the responsible development of nanotechnology, including, but not limited to—

(1) self-replicating nanoscale machines or devices;

(2) the release of such machines in natural environments;

(3) encryption;

(4) the development of defensive technologies;

(5) the use of nanotechnology in the enhancement of human intelligence; and

(6) the use of nanotechnology in developing artificial intelligence.

(d) EVALUATION TO BE TRANSMITTED TO CONGRESS- The Director of the National Nanotechnology Coordination Office shall transmit the results of any evaluation for which it made arrangements under subsection (a) to the Advisory Panel, the Senate Committee on Commerce, Science, and Transportation and the House of Representatives Committee on Science upon receipt. The first such evaluation shall be transmitted no later than June 10, 2005, with subsequent evaluations transmitted to the Committees every 3 years thereafter.

(a) NATIONAL SCIENCE FOUNDATION- There are authorized to be appropriated to the Director of the National Science Foundation to carry out the Director's responsibilities under this Act—

(1) $385,000,000 for fiscal year 2005;

(2) $424,000,000 for fiscal year 2006;

(3) $449,000,000 for fiscal year 2007; and

(4) $476,000,000 for fiscal year 2008.

(b) DEPARTMENT OF ENERGY- There are authorized to be appropriated to the Secretary of Energy to carry out the Secretary's responsibilities under this Act—

(1) $317,000,000 for fiscal year 2005;

(2) $347,000,000 for fiscal year 2006;

(3) $380,000,000 for fiscal year 2007; and

(4) $415,000,000 for fiscal year 2008.

(c) NATIONAL AERONAUTICS AND SPACE ADMINISTRATION- There are authorized to be appropriated to the Administrator of the National Aeronautics and Space Administration to carry out the Administrator's responsibilities under this Act—

(1) $34,100,000 for fiscal year 2005;

(2) $37,500,000 for fiscal year 2006;

(3) $40,000,000 for fiscal year 2007; and

(4) $42,300,000 for fiscal year 2008.

(d) NATIONAL INSTITUTE OF STANDARDS AND TECHNOLOGY- There are authorized to be appropriated to the Director of the National Institute of Standards and Technology to carry out the Director's responsibilities under this Act—

(1) $68,200,000 for fiscal year 2005;

(2) $75,000,000 for fiscal year 2006;

(3) $80,000,000 for fiscal year 2007; and

(4) $84,000,000 for fiscal year 2008.

(e) ENVIRONMENTAL PROTECTION AGENCY- There are authorized to be appropriated to the Administrator of the Environmental Protection Agency to carry out the Administrator's responsibilities under this Act—

(1) $5,500,000 for fiscal year 2005;

(2) $6,050,000 for fiscal year 2006;

(3) $6,413,000 for fiscal year 2007; and

(4) $6,800,000 for fiscal year 2008.

Section 7: Department of Commerce Programs

(a) NIST PROGRAMS- The Director of the National Institute of Standards and Technology shall—

(1) as part of the Program activities under section 2(b)(7), establish a program to conduct basic research on issues related to the development and manufacture of nanotechnology, including metrology; reliability and quality assurance; processes control; and manufacturing best practices; and

(2) utilize the Manufacturing Extension Partnership program to the extent possible to ensure that the research conducted under paragraph (1) reaches small- and medium-sized manufacturing companies.

(b) CLEARINGHOUSE- The Secretary of Commerce or his designee, in consultation with the National Nanotechnology Coordination Office and, to the extent possible, utilizing resources at the National Technical Information Service, shall establish a clearinghouse of information related to commercialization of nanotechnology research, including information relating to activities by regional, State, and local commercial nanotechnology initiatives; transition of research, technologies, and concepts from Federal nanotechnology research and development programs into commercial and military products; best practices by government, universities and private sector laboratories transitioning technology to commercial use; examples of ways to overcome barriers and challenges to technology deployment; and use of manufacturing infrastructure and workforce.

(a) RESEARCH CONSORTIA-

(1) DEPARTMENT OF ENERGY PROGRAM- The Secretary of Energy shall establish a program to support, on a merit-reviewed and competitive basis, consortia to conduct interdisciplinary nanotechnology research and development designed to integrate newly developed nanotechnology and microfluidic tools with systems biology and molecular imaging.

(2) AUTHORIZATION OF APPROPRIATIONS- Of the sums authorized for the Department of Energy under section 6(b), $25,000,000 shall be used for each fiscal year 2005 through 2008 to carry out this section. Of these amounts, not less than $10,000,000 shall be provided to at least 1 consortium for each fiscal year.

(b) RESEARCH CENTERS AND MAJOR INSTRUMENTATION- The Secretary of Energy shall carry out projects to develop, plan, construct, acquire, operate, or support special equipment, instrumentation, or facilities for investigators conducting research and development in nanotechnology.

(a) AMERICAN NANOTECHNOLOGY PREPAREDNESS CENTER- The Program shall provide for the establishment, on a merit-reviewed and competitive basis, of an American Nanotechnology Preparedness Center which shall—

(1) conduct, coordinate, collect, and disseminate studies on the societal, ethical, environmental, educational, legal, and workforce implications of nanotechnology; and

(2) identify anticipated issues related to the responsible research, development, and application of nanotechnology, as well as provide recommendations for preventing or addressing such issues.

(b) CENTER FOR NANOMATERIALS MANUFACTURING- The Program shall provide for the establishment, on a merit reviewed and competitive basis, of a center to—

(1) encourage, conduct, coordinate, commission, collect, and disseminate research on new manufacturing technologies for materials, devices, and systems with new combinations of characteristics, such as, but not limited to, strength, toughness, density, conductivity, flame resistance, and membrane separation characteristics; and

(2) develop mechanisms to transfer such manufacturing technologies to United States industries.

(c) REPORTS- The Council, through the Director of the National Nanotechnology Coordination Office, shall submit to the Senate Committee on Commerce, Science, and Transportation and the House of Representatives Committee on Science—

(1) within 6 months after the date of enactment of this Act, a report identifying which agency shall be the lead agency and which other agencies, if any, will be responsible for establishing the Centers described in this section; and

(2) within 18 months after the date of enactment of this Act, a report describing how the Centers described in this section have been established.

In this Act:

(1) ADVISORY PANEL- The term 'Advisory Panel' means the President's National Nanotechnology Advisory Panel established or designated under section 4.

(2) NANOTECHNOLOGY- The term 'nanotechnology' means the science and technology that will enable one to understand, measure, manipulate, and manufacture at the atomic, molecular, and supramolecular levels, aimed at creating materials, devices, and systems with fundamentally new molecular organization, properties, and functions.

(3) PROGRAM- The term 'Program' means the National Nanotechnology Program established under section 2.

(4) COUNCIL- The term 'Council' means the National Science and Technology Council or an appropriate subgroup designated by the Council under section 2(c).

(5) ADVANCED TECHNOLOGY USER FACILITY- The term 'advanced technology user facility' means a nanotechnology research and development facility supported, in whole or in part, by Federal funds that is open to all United States researchers on a competitive, merit-reviewed basis.

(6) PROGRAM COMPONENT AREA- The term 'program component area' means a major subject area established under section 2(c)(2) under which is grouped related individual projects and activities carried out under the Program.

Focus On Social, Ethical, Legal, International And National Security Implications

Scientific discoveries by themselves rarely create change. It is the confluence of old and new technologies with old and emerging social needs that creates change. There are dynamic interactions and "tipping points" that are difficult to foresee (e.g., the capabilities that emerge when computers get one more order of magnitude faster, coupled with a clever new idea for software, leveraged by increasingly computer-literate 20-yearolds).

Thus, the problem for prediction is not only to track the advances of nanotechnology, but also to track other advances and changes in society at the same time. Visionaries and science fiction writers have bold visions, but researchers must also take a role in producing a useful understanding of possible societal directions.

Perhaps the best we can do is to open our thinking and be more aware of the kinds of changes that may occur, how to spot them as early as possible, and how to prepare to influence the course of changes when deemed necessary or desirable. Of course, it is important to consider who would be doing the "deeming." History suggests that those in power tend to suppress or co-opt new technologies (e.g., RCA's success in subverting Farnsworth's patents on TV and convincing the public that they were the inventor, Fisher and Fisher 1997), and the newly empowered try to do the same (cf., Bill Gates).

So, I accept my more modest role to suggest social science research methods for studying societal implications. I begin with some assumptions about research and nanotechnology, and then present goals for social science approaches to societal implications of nanotechnology. I then suggest sources of measures and kinds of indicators that can be helpful, including "leading indicators" that might provide early hints of change. Finally, I offer possibilities for types of research designs and some conclusions.

Nanotechnology, as a family of tools and techniques, is a source of products and other "stuff." It is these products and stuff that will be fought over, demanded, and used in ways that will bring layers of change. Each of these bits has the potential to be mundane variations with a bit more zip or a bit less cost, or to be dramatic advances. We are not likely to know which is which, and the nanotech equivalent of the hoola hoop may turn out to be critically important when combined with something else.

Initially, the impacts of nanotechnology will be via specific products and innovations. Such primary effects would be to make things work better, cheaper, with more features, etc. This might, for example, increase food yields, generate new textiles for clothing, improve power production, cure certain diseases, or whatever. Secondary effects might be shifts in demand for products and services, so that people come to expect different kinds of food, medical care, entertainment, etc. The required infrastructure for nanotechnology may create interdisciplinary research centers, new educational programs to supply nanoscientists and nanotechnologists, etc. Later, tertiary effects would move upstream in our social structures and cultural patterns, such as shifts in education and career patterns, family life, government structure, and so forth1. Will nanotechnology extend our lifetimes more rapidly than it extends our health, or vice versa? Will nanotechnology enable so much connectivity to information and to each other that it revolutionizes society (cf., Star Trek's The Borg) or shift us more toward solitary lifestyles of "distance experiences" and "virtual experiences" rather than personal contacts (cf., Asimov's *The Naked Sun*)? Research, including social science research, is an activity intended to generate and validate knowledge, based on systematic rules agreed to by a community of scientists.

Since there are many subcommunities of scientists, the rules vary somewhat from discipline to discipline, place to place, and time to time. It is important to realize that the activities of social science research also influence policymakers and the general public, and thereby change the way society thinks and acts (Gergen 1973). To measure something is also to shape the future, by changing what we pay attention to, expect, reward, and punish. Therefore, it is important to measure both what is desired, and also what is feared.

In light of the above assumptions, it seems reasonable to pursue the following goals for social science research on the societal impacts of nanotechnology. First, we need to define and measure "societal impacts." Second, we need to find leading indicators or first signs of impacts. Third, we need to develop theories that explain impacts, identify causal mechanisms and contingent conditions (e.g., under what circumstances would particular products have particular impacts), relate various advances and impacts together in more comprehensive systems models, and permit (tentative) extrapolation to possible futures. Finally, we would like to assist policy development on the basis of what is known from our research and what is known about desires and values, i.e., what are "society's" goals and how will these goals change over time as technology advances?

There are a huge number of potential measures for societal impacts. Our standard measures of the social world are all useful and all insufficient. From a methodological viewpoint, there are various techniques for measuring societal phenomena, including selfreport surveys and interviews; diary studies (e.g., new software that allows Palm Pilots to interview their owners) and think-aloud protocols; social and economic statistics; direct observation (of, for example, meetings, point of sale, point of use); bibliometric and content analysis (of patents, citations, email, Web sites and hits, chat room content, 10Ks); network analysis of relationships across people and organizations; accretion measures of what piles up (e.g., trash heaps); and erosion measures of what gets worn out (e.g., repair calls). For some general principles and examples, see Babbie 1989; Judd et al. 1991; Webb et al. 1981).

It may be helpful to sort through potential measures by using some conceptual criteria or categories. For example, we could distinguish *process* from *content*. The temporal *process* might include scientific or social visionaries who first imagine possibilities, nanoscience and nanotechnology discoveries that connect with these possibilities, actual products that are developed and marketed, public acceptance by scientists, consumers, and policymakers, substitutions of usage patterns as new products replace old products, interactions of new product capabilities with existing technical and social arrangements that reshape demand and usage, and finally transformations of social institutions and associated infrastructures. The *content* domains might include health, wealth, food, fuel, productivity, education, employment, national security, happiness, social capital (networks of relationships), political participation, ethical thought, etc.

From the process categories, we can generate more specific indicators of innovation and change. Visions are likely to appear in the media, at meetings, and on the internet. Discoveries can be tracked through patent applications and new products and services. Investments are visible in budgets, grants, strategic plans, projects, job titles, hiring results, and educational programs. Public acceptance is indicated by attitude measures, purchases, usage, and even charitable contributions. Interactions show up in studies of professional and other social networks, fields of study, and communities. Substitutions are evident in the decline of old industries. Transformations can be measured in indicators of lifestyle and social institutions such as deurbanization and demassification .

Other conceptual criteria underlie the objectives and recommendations in the IWGN Workshop report. These constitute a nascent "theory" of the nanotechnology process and impacts. The report divides its objectives and recommendations into sectors or classes of participants such as academe, private sector, government labs, funding agencies, and professions. For each sector, the objectives suggest process and/ or content measures of change. In academe, the workshop endorsed interdisciplinary work, new courses, fellowships, information flow, and regional coalitions. For the private sector, the focus was on investments, startups, and coalitions. For government labs, the report looked at budgets, equipment, standards, and coalitions. For funding agencies, the priorities were on new initiatives, databases, and centers. The professions were expected to create new forums, symposia, and job fairs where interdisciplinary topics and careers could flourish.

Of particular importance would be indicators that provide early signs of change. In our process indicators, we would probably look upstream to the beginning of the chain of events and examine allocations of effort in academic and corporate R&D labs and analyses of patent applications to see the emerging technological trends. More generally, we might consider who, what, and where we could find "bellwethers" or first movers.

California seems to be the first place that social change occurs in the United States. Finland and Singapore are examples of countries that have embraced new technologies and undergone rapid change. Within both rapidly changing and slowly changing societies, there may be classes of "early changers" such as 15 year olds who are at the forefront. In some cases, such as in the health care domain, it may be the 90-year-olds who show the first signs of change. Science enthusiasts, nerds, and hackers may be first movers in a nanotechnology-rich world. Start-up companies, university labs, and Internet chat rooms may be places to look for changes and impacts. Particular social strata, such as groups "on the margin" of society, may also exhibit the early signs of changes (things that seem weird or bad) that will later spread to mainstream society.

Research is more effective when it is designed into the process being studied, rather than having to explain what happened after the fact. It is not surprising that many of our best studies of the impact of technologies are retrospective analyses from 10 or 100 years ago (e.g., Bijker 1995; Fischer 1992) but we can't afford to wait that long for knowledge of current developments and impacts. If social scientists can be introduced earlier into partnerships and collaborations with nanoscientists and nanotechnologists, there is a better chance to learn more and learn quickly enough to guide policy. Research designs are typically stronger on measures and mechanisms if introduced early, so that measures can be made over time and informative controls can be established to strengthen causal understanding.

I have listed some typical design types in the order in which I would guess they will be used. In other words, designs that are easier to execute are likely to be used first, and those that require great skill and/or great control over circumstances may come later.

Surveys, expert panels, and statistical summaries of socioeconomic data are relatively easy to do and provide useful snapshots. Simulations (computer-based or game-like) allow us to project behavior in hypothetical scenarios (e.g., Sterman, 1989). Case studies (Yin 1989) permit a rich set of information gathered around particular cases of inventions, products, companies, communities, supply chains, etc. Quasi-experiments (Cook and Campbell 1979) allow focused comparisons that strengthen causal inference, such as time series and econometric designs and control groups based upon diffusions of new technologies (some groups naturally get it earlier, some later). Ethnographies (Van Maanen 1988) and in-depth immersion in real-world sites allow a rich understanding of complex interdependencies and subtle phenomena. True experiments allow us to answer precise questions, but are difficult to arrange in real-world settings.

In conclusion, we must remember that we cannot easily predict the products and impacts of nanotechnology on society. We are not studying a monolithic "nanotechnology" but rather a host of varied technologies, products, services, and other interventions. Some implications and impacts are relatively easy to predict and to study, but others will be emergent and surprising. However, everything in society will be changing (not just what nanotechnology has touched directly), and everything will be connected together (loosely or tightly), so predictions will be uncertain and causal explanations will be difficult to validate.

Research is needed to help us understand changes and to plan action. A wide range of indicators will be needed because we do not know what will emerge or what will turn out to be important. Helpful theories are especially important to focus attention on key issues and processes, to guide research, and to represent the results of research. Society needs theories and system models to understand how changes in one part of the system, whether a particular type of technology or a particular element of society, spread out to create intended and unintended effects throughout the system. This includes understanding the impact of society on technology. This will enhance our ability to educate everyone earlier and to improve the quality of public debate. It will also enable scenario analyses, strategic planning, and simulated public debates to be more informative and rich.

At its best, the research attitude is one of openness, curiosity, sharing, and constant improvement. This is a model for increasing the capability of the public and the scientific communities to plan intelligently, communicate effectively, respond to emergent circumstances, and understand themselves and the broader society and world in which they live and work. From an ecological perspective, there is no guarantee that "progress" (however defined) has any particular consequences for the human race; nor is there a guarantee that more science and technology will always find an answer to human problems. We must improve ourselves as a thoughtful and ethical human society at the same time that we improve our mastery over the physical world.

Because we are concerned with issues of practical and professional ethics, not theoretical ethics, we must examine nano initiatives along with associated social practices, institutions, organizations, and the choices and actions of individuals within them. In nano science and engineering, ethical and responsibility issues connected with various initiatives will reflect many complexities — those from the multidisciplinary and multiinstitutional character of research and development, as well as from more technical aspects. To survey ethical implications and identify risks it is necessary to step back from captivating visions of profound transformations of the material world and of society. The focus must be on specific proposed initiatives in their institutional environments. Normative questions, that is, questions about what it is right and appropriate to do, gain a foothold at the macro level in relation to societal and organizational policies and activities and, at the micro level, in relation to ethical standards, responsibilities, decisions, and actions of individuals. An important aim of ethical investigation is to anticipate ethical problems — preventable harms, conflicts about justice and fairness, and issues concerning respect for persons likely to arise from specific nano initiatives. A second important aim is to foster sensitivity to ethical issues and responsibilities at every level of decision making by both technical and policy people.

At the outset, it is essential to clarify the term "nanotechnology." There is no disagreement about the oversimplifying and misleading character of that term. It is a catch all that has caught on because of its convenience and market appeal, its usefulness as a rallying point. Though aware of its deficiencies, interested parties are not inclined to reject use of the term. It is therefore necessary to make clear what are the initiatives, disciplines, and institutions that "nanotechnology" embraces.

(1) Initiatives with implications for advances in medicine, computing, space exploration, energy conversion and storage, optics, and materials, including catalysts, are among those often cited.

(2) A large number of disciplines — and specialties within them — in science, engineering, mathematics, and computing are encompassed.

(3) Institutions of government, academe, and industry, and organizations and practices within these institutions are also included.

Government agencies act as promoters of initiatives, and they have begun to fund university research and graduate training. Universities are gearing up for and launching new programs, instituting graduate studies, renovating and putting up buildings, and forming alliances with other universities, national laboratories, and private firms. New collaborative arrangements between universities and private companies to carry forward research and development are being, and will continue to be, forged. Of use in forming these new collaborations is the experience recently gained from crafting such arrangements to advance work in information technology and biotechnology.

That experience suggests a second task: to examine recent history. Study of our experience with biotechnology and information technology may help to locate nodes of ethical concern. Caution in drawing parallels is necessary, however, in light of claims made for the uniqueness of nanoscience and nanotechnology. It may, nevertheless, be useful to direct attention, from an ethical perspective, to questions that these earlier technologies suggest. Included are questions about how government plays a role in promoting, launching, and supporting technological developments and about how projects to produce products are selected and by whom. Government agencies' promotional discourse to boost the NNI is already available for scrutiny, and there may be opportunities to study how nano projects are currently selected. Our history with earlier technologies suggests the need to devise processes and settings for information exchange with and wider participation by members of the public in order to promote transparency. (See Wynne 1991 for an especially insightful discussion.) So compelling are the ethical and practical benefits of building in openness, disclosure, and public participation from the outset that efforts towards those ends should begin without delay.

Recent history with other technologies indicates that the obstacles to achieving these aims are formidable. Observers have already expressed concern that multidisciplinary meetings present special risks to open exchange of information (NSTC 2000, p. 32). Similar concern is appropriate regarding the flow of information across institutional boundaries, between academe, government, and the private sector (Blumenthal 1992). Some problematic patterns relating to the flow of information between those engaged in technological development and the public are familiar from technological developments of the recent past. For example, by becoming locked into a definition of the public as the "other" — the enemy, uncomprehending, standing in the way of advance — those with authority over information may withhold it and thereby cut themselves off from public reactions. Failing to nourish genuine information exchange, they may invite the very opposition they wish to ward off. Such patterns from the past point to the need for care and caution in framing "the communication problem", to avoid seeing it as a problem of one-way communication downward (Wynne 1991).

A third task at this stage is to try to identify ethical issues that have already arisen or are likely to arise. As elsewhere in practical and professional ethics, it is necessary to disaggregate, to look case by case at specific nano science and technology options and their consequences — current, foreseen, foreseeable, or speculative. Initiatives and options should not be divorced from their institutional contexts, however complex the latter may be. At times technical people fail to foresee what is foreseeable and within their sphere of responsibility. Sometimes they claim to foresee what is not foreseeable, given available knowledge. Nanotechnology options afford opportunities for both kinds of foreseeability problems to arise regarding the consequences of specific developments.

The recent history of rapidly developing technologies suggests that we should be alert to unintended consequences. Consider, for instance, the complex issues that have arisen about privacy in connection with information technologies (Johnson and Nissenbaum 1995). The privacy debates also illustrate how ethical issues concerning respect for persons are generated with the propagation of new technologies. Biotechnology offers another cautionary example relating to unintended consequences. When investigators in the new biotechnology achieved confidence in control of their products in the laboratory, vigilance regarding unintended consequences did not extend to the new products in the field. Information about interactions of new bioengineered organisms with other members of an actual ecosystem has been scarce and slow to appear. Accordingly, concern about unintended consequences from agricultural products of biotechnology remains high in many places (Weil 1996).

While trying to stay alert to unintended consequences, we should also try to avoid taking it for granted that there is wide agreement on the desirable consequences of various nanotechnology options. It is essential to obtain a diversity of perspectives on the desirability of particular options. For eliciting a range of perceptions, conversation between people in nano research and development and members of the public is necessary. It may be essential to create mechanisms, such as citizen/scientist panels, to bring different perspectives into the conversation. In the exchanges, people will have to learn to identify interests that are in play, their own and those of other parties to the conversation. Experience with biotechnology shows the costliness of proceeding with mistaken assumptions about what are desirable outcomes and products (Crow 2001).

To go further in identifying ethical issues, we need concrete points of departure, actual examples or cases that pose ethical questions, quandaries, or conflicts. This is how we proceed in practical and professional ethics when looking at issues in other social practices. Lacking specific nano examples at this time, we may note features of nanotechnology that, because of their alleged novelty, may be sources of ethical concern. For example, forecasts of development of new catalysts stress the creation of new production processes. The latter are likely to raise ethical questions, for instance, about the need for safeguards for workers that specific new processes might generate. Questions of these kinds can provide points of entry to the institutional, organizational settings in which potential problems are embedded and in which they must be examined. The new processes are confidently predicted to produce workforce changes (NSTC 2000, 20). Anticipated impacts on the "human resource infrastructure" will surely bring benefits to some and harms to others. When there are winners and losers, issues about equity cannot be avoided. Again, on a case by case basis, it will be important to identify winners and losers so as not to inflict preventable harms or lose opportunities for mitigating harms or compensating losers.

Ethical issues associated with intellectual property protection are virtually certain to arise (NSTC 2000, 31). They will be novel insofar as novel features of nanotechnologies and their social environments introduce new complexities relating to intellectual property protection. In environments where patents and trade secrets are generated, there will be implications for open exchange among technical people and communication with the public. The rationales justifying ownership are likely to be as vigorously debated and contested as those associated with information technology and biotechnology. Equity issues raised by intellectual property protection should generate debate as well. Vigorous and extensive public discussion could even lead to reexamination and revisions of intellectual property policies.

Ethical questions about university/industry relationships are hardly novel, but they are virtually certain to arise. These questions have engendered discussion, literature, and large-scale empirical studies since at least the mid 1980s (Blumenthal et al. 1986). By now, some specialists contend that institutional accommodations to new relationships with private companies have transformed universities, bringing significant changes in university values and practices (Webster and Etzkowitz 1991). For instance, many in universities now accept a need to allow faculty members to accumulate great wealth through their research (Forest 2000). Yet there is a clear understanding on both sides that universities and private sector enterprises are valuable to each other as partners precisely because of their differences (Weil 1988; Weil 2000).

The close association of university research with the private sector has brought problems of conflict of interest to the forefront. For example, questions arise about whether a university researcher's ties to a for-profit firm threaten reliable judgment in university research. Observers have suggested that universities as institutions can have conflicts of interests (Frankel 1996; Pritchard 1996). A strong program of research and development in nanoscience and nanotechnology will subject university values and practices to new pressures, and universities will have to continue to make accommodations preserving core university values. They will have great need for ethical guidelines as they make those accommodations and review their policies in the light of experience.

The focus on institutions, organizations, and practices should not obscure the need to focus on the individual responsibility of engineers, scientists, and others involved in the processes of producing new technologies "with unprecedented control over the material world". To focus on individual responsibility, university programs of graduate study and research in nano areas should include attention to ethical issues specific to their own nano areas. They should give attention to these issues in their training of students in scientific research ethics and in their technical training of students. Professional societies have a role to play in affording opportunities for debate and discussion and helping to devise, for individuals and organizations, guidelines that incorporate ethical principles responsive to emerging issues. This survey of ethical dimensions suggests three main fronts of activity for responding to issues. One is activity within research and development initiatives funded by government agencies and carried out in universities. Government agencies should require an ethics component in grants for graduate training and in grants for research. In this way, they can provide for attention to ethical issues by individual researchers and principal investigators within research groups, by research groups as collectives, and by university departments and centers in which the nano research and training goes on. It is necessary to connect specialists in ethics and behavioral sciences with such projects from their outset and maintain the specialists' association with ongoing nano projects of research and development. National Science Foundation proposal guidelines already incorporate provisions requesting an ethics component. Such provisions should be standard components of requirements for submitting proposals in the nano domain to government agencies. Over time, the provisions can be further detailed or improved in other ways, as experience indicates.

The second front of activity has already been suggested — a fresh and energetic effort to devise opportunities for genuine conversation with members of the public about current and proposed initiatives. In view of the ethical and practical reasons for commitment to disclosure and incorporation of democratic processes in advancing nano initiatives, government agencies that are promoting nanoscience and nanotechnology should accept an obligation to create appropriate channels and fora. Concerted activity on these two fronts could be innovative and capable of transforming research and development in step with the innovative and transforming features of nano processes, structures, and products.

A third front of activity is education. While some ethics specialists are already available to help in addressing the ethics component in research, development, and education related to nano initiatives, there are not enough. Mainstream graduate training in philosophy and behavioral science is not yet oriented to respond to this need. The intellectual interest and acknowledged importance of issues raised by the NNI justifies government initiatives to address the lack of qualified ethics and behavioral science specialists. It should be possible to devise training programs led by existing specialists in collaboration with faculty who direct graduate and postgraduate study in appropriate disciplines. Postdoctoral studies may be especially suitable for providing the training and forming the collaborations across disciplines that are needed. A program to train people for the ethics work within nanotechnology and nanoscience initiatives is essential, and feasible, with government support. This education should be part of a concerted endeavor to improve education at all levels, in scientific and engineering disciplines, and areas that cross disciplines, as well as in philosophy, ethics, and behavioral sciences. For the NNI to begin to fulfill its promise of carrying forward research and development initiatives that bring benefits to society fairly distributed, concerted efforts on all three fronts are needed.

The notion of "social acceptance" of technology is prevalent throughout both scholarly and social science studies of technology and in popular literature, yet it is, in an obvious way, a very obscure idea. On the one hand, it connotes empirical content, perhaps even measurable criteria, so that whether or not a technology has been socially accepted appears to be decidable question, a matter of fact about social relations or how things stand in the world. On the other hand, the phrase "social acceptability" suggests a normative judgment in a way that makes social acceptance come to involve inherently contentious characterizations of "society's values." Here I want to make a few general remarks about the social acceptance of new technology, to note some lessons learned and yet to be learned from the ongoing saga of biotechnology, and to make some suggestions about how these lessons might apply to nanotechnology.

Here are some of the indices that might be taken to measure social acceptance of technology.

- *Geographical Parameters*: Where is the technology used — and "where" can be characterized in terms of location, demography, culture, class, etc.
- *Economic Parameters*: What is the market penetration for the technology (what percentage of the potential purchasers of the technology actually purchase it), what is the price sensitivity, etc.
- *Psycho-Social Parameters*: What do surveys indicate when people are asked how what they think about a given technology?
- *Affective Parameters*: What's the "comfort level" of users? Do they feel a sense of regret? Do they feel a sense of moral disapproval? (One could argue, for example, that despite widespread use, chemical pesticides are not a socially accepted technology).
- *Cognitive Parameters*: What's the level of awareness that a technology is being used? Are people presented with clear opportunities to accept or reject a technology? (Arguably, many of the technologies in daily use have not been "socially accepted" simply because they are, for the most part, wholly unknown.)
- *Technical Administrative Parameters:* Where is a given technology in the process of regulatory review and approval? Are there regulatory or court decisions that actively sanction the use of the technology?
- *Political Parameters:* What is the level of debate, contention and organized opposition to a technology? What is the potential to mobilize opposition at any given time, and at what cost?

There are also normative and quasi-normative parameters for the social acceptability of technology. Here are a few:

- *Religious Acceptability*: What do religious doctrines or religious authorities have to say about the technology? There are cases (such as whether cheese made with recombinant chymosin meets Kosher standards) where these are rather straightforward questions.
- *Cultural Acceptability*: Given the fact that many cultural norms are implicit and veiled, the question of whether a given technology is culturally acceptable is often vexed and contentious.
- *Ethical Acceptability*: I take this to be the broadest and most irreducibly normative category. *Should* society accept a given technology, and if so, under what constraints or qualifying conditions? (So, one might argue

that a technology is acceptable only if its advantages and benefits will be distributed fairly across all economic classes).

Many of the social sciences employ forms of analysis that generate evaluations that are at least hypothetically ethical, in the above sense. Standard economic cost-benefit analysis is an obvious example. Though saying whether the costs of a technology outweigh its benefits is not a normative judgement in the sense of actually recommending for or against, such analyses nevertheless suggest qualified normative criteria that might be (and indeed commonly are) used for such a judgement. I hope it goes without saying that what "cost," "benefit" or "fairly" means is itself an ethical issue, that what it means for society to accept a technology (in a normative sense) is an ethical issue, and that whether there even *are* such normative criteria is an ethical issue. One can debate terminology, but the category of ethical acceptability is logically inescapable, since to deny the relevance or possibility of ethical criteria is to make a contestable normative claim. Criteria for descriptive parameters for social acceptance tend to become entangled with normative criteria for social acceptability. The reasons have little to do with the study of technology *per se*. For example, behavioral social science may require framing assumptions about rationality in order to structure data bases or to characterize a behavioral phenomenon as an instance of "choice". But calling a form of judgment or behavior *irrational* is generally taken to imply a normative judgment about it. This opens out into both philosophical and highly tendentious debates, especially about the kinds of social science analysis noted above. While it is tempting to dismiss talk about the social acceptability of technology and society's values as lacking proper rigor (Is society really the sort of thing that can have values?), it is useful to recognize some important things that are going on in such talk.

Clearly, players — and we are all players here — try to "spin" the social acceptance of a technology such as nuclear power, food irradiation, e-commerce, or biotechnology in an effort to influence opinion. That is, social acceptance is a thoroughly reflexive phenomenon. Given the multiple ways (above) that a technology might or might not be characterized as socially accepted (or not), it is generally possible for those who wish to quiet or aggravate acceptance as measured by any given parameter to say (with some degree of truth) that the technology is/is not/may/may not be/will be/won't be socially accepted (with respect to some other parameter).

Given the economic, political and cultural and professional interests that are (or might potentially be) at stake, anyone may be an interested party. This is *not* to say that there are no facts with respect to the indices that might be taken to measure social acceptance of nanotechnology, nor to say that everyone wants to influence the social acceptance of a technology that they study. But it does suggest that any reported observation regarding these indices may with fairness be subjected to a normative critique. For my money, the reports of sources that disclose their interests will be more credible.

Most scientists (including social scientists) are not trained at clearly understanding their own interests, much less at disclosing them openly to others. Unfortunately, the arguments that are proffered in enduring philosophical debates, edifying in their own right and crucial to rigorous social science, are quite likely to be deployed in the strategic gamesmanship of players seeking to advance or retard the social acceptance of a given technology. Has biotechnology been socially accepted? I will confine myself to food and agricultural biotechnology, which is my primary area of expertise. Certainly there are published reports that bear on each of the indices noted above.

- *Geographical Parameters*: One can measure the number of hectares sown in biotech crops, the countries in which they are grown, and the demographic characteristics of farmers who use biotech seeds.
- *Economic Parameters*: There are many economic studies of the adoption of biotech, as well as studies of the competitiveness of the industry.
- *Psycho-Social Parameters*: There have been repeated surveys of public opinion on biotech. Levels of acceptability vary in degree and in change of direction across national groups.
- *Affective Parameters*: Surveys and focus group research provide some indication of affective parameters. This is an ongoing area of data collection and qualitative analysis.
- *Cognitive Parameters*: Surveys indicate a general measure of awareness, generally quite low, it is worth noting, for a technology that has become a model case study for a contested technology.
- *Technical Administrative Parameters*: Since the early nineties there have been regular reports of regulatory approvals. The weakness in this area would appear to be the lack of a general theory of what might have been thought to be an obvious question: what constitutes legal or administrative acceptance?
- *Political Parameters*: A substantial amount of qualitative research exists. There do not appear to be standard methodologies for developing indices of political contentiousness and acceptability.

Some Lessons: Though a fair amount can be said about each of these parameters as they relate to food and agricultural biotechnology, what do we know about the social acceptance of food and agricultural biotechnology? I would be skeptical of anyone who professed to know whether this technology *has been* accepted or *will be* accepted, both on a global or a regional basis. While it seems unlikely that agricultural biotechnology will simply disappear, it is anyone's guess as to whether the levels measured in these indices are stable, or will increase or decrease over time.

To me, this suggests that we don't know very much about how these parameters affect one another, about the dynamics of social acceptance. The notion of social acceptance seems intuitively clear, and even demonstrable for some historical cases. For example, can we really question whether electrification has been socially accepted in the industrialized world? Yet I would question whether we really have a very good sense of what real-time social acceptance amounts to. The strategic, normative and reflexive dimensions of social acceptance may account for the open-ended nature of the real-time index for social acceptance. But it would be useful to have a theoretically clear and rich statement of why this is the case.

Has the battle of strategic positioning for the social acceptance of nanotechnology already begun? It is inevitable that, whatever the motives of their authors, the documents which already exist (the documents we are generating at this workshop) will be spun by players down the road. Because I have ideas about how which criteria and procedures should lead to the acceptance, rejection or qualification of any new technology, I would argue that several things should be done at the earliest opportunity. Most importantly, science funders should invest in the creation of fora that are shielded from strategic actors to the extent possible. There are, of course, limits to the extent that this *is* possible, but several of the following things would help. One is clear, open and ongoing multi-disciplinary clarification of the interests that are advanced and retarded by the development of nanotechnology. A second goal would be to increase scientists' capacity to reflectively understand the sense in which they are interested parties, and encourage disclosure of those interests. Here, career, disciplinary and ego-based interests can be as decisive as pecuniary ones. One should not shy away from development of explicitly normative studies and position papers, but such efforts should aspire to high standards of transparency, clarity of analysis and to the creation of a public record.

It would also be useful to build on the literature of social acceptance as it has been developed with respect to technologies such as biotechnology, to design studies that would compare and integrate the indices described above, and to examine analogies that might suggest a basis for normative evaluations of the social acceptability of nanotechnology. The "technology out of control theme" or the precautionary principle as applied to biotechnology might, for example, present a suggestive starting point for normative studies on the acceptability of nanotechnologies.

The scientists and engineers behind nanotechnology must be involved in all the above, and to me this suggests that there will be a need for an ongoing series of conferences, workshops, seminars and publications with a fairly high level of visibility. Such people will exclude themselves from the standard run of social science disciplinary research outlets, so something else is needed. These activities need to be visible enough so that someone who gets interested in the ethics or social acceptance of nanotechnology fifteen years from now won't need extraordinary good luck to find the public record. That's sorely missing with respect to agricultural biotechnology, where the social acceptance wheel is constantly being reinvented by players and naïve researchers alike. Perhaps that points to a national center, or perhaps to a program like the Ethical, Legal, and Social Implications of Human Genetics Research (ELSI) at the National Institutes of Health.

The term "nanotechnology" may include such products as mono-layered materials, nanocomponents in smarter MEMS, and Fullerene-based computers. Some decades from now, it may also include communicating and/or programmable molecular machines. The implications discussed herein assume that some decades from now, the latter capabilities are achieved.

In the short-term, incremental improvements in processes and materials can come about by improved knowledge and skills in the nano-realm. The additive effects of these changes can have secondary and tertiary impacts that are transformational. For example, the social, ethical, and legal impacts of the World Wide Web were not originally thought to be considerable. The Web's enormous impacts are still not fully understood. In a similar way, nanotechnology may appear gradually and yet have a revolutionary effect. In the longer-term, the risks and rewards of nanosystems will certainly be exaggerated as our technological capabilities improve. Virtually every "millennium survey" of the future poses some social, ethical, or legal questions about nanotechnology. But there is little so far in the way of serious study. Bill Joy's recent article, "Why the Future Doesn't Need Us" in *Wired* is a widely read example. Mr. Joy suggests that since nanotechnology is potentially dangerous, we should relinquish our study of it. His article, though intriguing, misses a most critical point. The United States doesn't have a monopoly on nanotechnology research. The rest of the world is spending over $1 billion per year in the field. Relinquishment by friendly governments, even in the unlikely event that it could be enforced, does not ensure that all researchers will make the same decision.

Others have also encouraged the public's fears about nanotechnology and biotechnology. Just as fear of cloning could slow efforts in biomedicine and fear of genetically modified foods could contribute to hunger, fear of futuristic nanobots running amok could delay the benefits offered by nanotechnology. When discourse is founded on emotional prejudices, its unreasonableness discredits the legitimate need to identify and assess risks.

Nanotechnology could ultimately turn out to be risky, but the prudent way to assess the risks is not the abandonment of the field. Asilomar demonstrated that the scientific community could design systems to contain high-risk technologies; the Shelter Island meetings showed us we could put great minds towards thinking about risk; the space program showed us we could have a dangerous program with almost no fatal missions. Notwithstanding the negativists, we have proven that we can manage the risks of powerful technologies. This does not suggest that we are safe — merely that we are not inevitably doomed to the worst of possible technological outcomes. We should hear from the pessimists, but we should not hear *only* from them. Here are some examples of possible social, ethical, and legal implications that bear some thought and consideration. You have a handout that contains a more robust list.

Medium Term

- The NNI will fund careers for scientists and graduate students; commercial firms will invest in nanotechnology R&D. These efforts will be widely dispersed politically, geographically, and scientifically.
- Nanotechnology research will both require and produce enabling technologies that will have beneficial spin-offs.
- There will be unintended consequences, both good and bad.
- We will have to balance the opportunity costs of studying assembler-based nanotechnology (with its huge potential payoffs) against research in fields where advances might have nearer-term but smaller payoffs.
- Designs for future nanosystems will be created, challenged, modeled, improved.

Medium Term

- Nanotechnology research may allow an otherwise moribund Moore's Law to continue operating.
- Fullerene-based computer chips may require enormously expensive fabrication facilities that turn out chips by the millions — needed for smart packaging, foods, Bluetooth devices, etc. This could result in a proliferation of inexpensive unit costs but prohibitively (and anti-competitively) expensive initial capital costs.
- We may discover and perfect nano-sized sensors and tools that can diagnose disease much sooner than ever before — perhaps long before we have cures.
- We should expect revolutionary advances in materials, MEMS, etc., resulting in abundant markets but also in an upheaval in global financial and manufacturing systems.
- Public education may be needed to balance the views expressed by anti-technology writers and press.

Long Term

- Nanosystems may help solve problems of disease and aging, pollution and scarcity, overpopulation and starvation, and could create revolutionary changes unlike any ever seen.

- Nanosystems could help produce alternatives to fossil fuels and their high environmental price and reliance on foreign sources.
- The transition from a pre-nano to a post-nano world could be very traumatic and could exacerbate the problem of haves vs. have-nots. Have-nots do not easily obtain access to new technologies; the difference between the lives of the nano-rich and the nano-poor will likely be striking.
- Potential harmful uses — intentional and unintentional — need to be studied well in advance: nano weapons; intelligence-gathering devices; nanotechnology combined with Artificial Intelligence to form super-intelligent but virtually invisible devices; artificial viruses to which humans have no immunity, etc.

Recommended Social and Ethical Research Areas

- We should follow a broad path in the ethical studies of nanotechnology including utilitarian ethics (in its many forms), virtue ethics, communitarian ethics, deontological and religious studies, and views of administrative and distributive justice.
- We need to look at the broadest reasonable set of worldviews if we are to do justice to the problem.
- Scientists and engineers like those at Rice University, New York University, MITRE Corporation, and DARPA can inform us on what is feasible.
- Technology pessimists can probe areas of risk that might escape a less vigorous review.
- Optimists from organizations like the Foresight Institute can provide insights into the kinds of systems that seem to be within the realm of the possible if assemblers can be made to work.
- Organization and process professionals like those from ITRI and the Institute for Alternative Futures can coordinate experts who would typically not relate to each other.
- Technology assessment professionals like those with Coates & Jarratt, Inc. can apply time-tested tools to estimate a range of possible circumstances and assess the secondary and tertiary effects of incremental and radical changes.
- We need formal risk analysis because risk is one of the most important ethical and social issues we could imagine:
  - Risk of physical harm—Risk to economic and social systems
  - Risk to political and financial power bases

- Risk of the haves fighting to keep the upper hand over the have-nots

- We need experts from several legal disciplines to understand the legal implications:
  - Litigators
  - Patent attorneys
  - Privacy law specialists
  - Constitutional lawyers
  - Academic legal scholars
  - Attorneys who understand the process of using the law as a weapon to retard progress

Most importantly, a multidisciplinary approach is critical to a satisfactory understanding of the social, ethical, and legal implications of nanotechnology. For this reason, there needs to be a gathering of all the PIs and other listed key personnel from each grant at least once every six to nine months with mandatory attendance and at least one paper presentation by each sponsored grantee institution (firm, individual, or university). This will not be inexpensive or easy to administer, but it will increase the odds of researchers understanding the issues.

At some point over the course of the next several years, we will have to think about how — scientifically — we are going to achieve a nanotechnology that includes communicating or computing capability. In order to assess the unintended consequences — secondary, tertiary, quaternary uses and effects of new technologies — we must consider what approaches are used. Otherwise, we can't possibly think forward to what unintended consequences might be or how they might be manifested. Nanotechnology arrived at through chemistry may have quite different characteristics than nanotechnology arrived at through biology.

In the spirit of Bonnie Nardi's paper in this volume (Nardi 2001), I would like to "envision" some of the legal and organizational challenges that might arise from the widespread introduction of nanotechnology into the U.S. economy. Obviously, true prediction currently lies beyond the capabilities of even the most informed and foresightful experts in the field, let alone an outside observer like myself. Nonetheless, social scientific research on previous technological revolutions suggests at least a few tentative hypotheses. In particular, as we speculate on policy implications, we might be well advised to distinguish between two types of nanotechnology — which I will call nano-materials and nano-mechanisms, respectively — and to explore the implications of each separately. I suspect that the social disruptions and governance challenges stemming from the former are likely to be much more localized and manageable than those stemming from the latter.

Accounts of the coming nanotechnology revolution seem to involve two somewhat separable agendas — each with its own distinctive social implications. The first (and, in the short run, probably more significant) agenda links nanotechnology to chemical engineering and material science. This variety of nanotechnology focuses on controlling the nano-scale organization of macro-scale substances. Examples of such "nanomaterials" — or "nanates" — might include wear-resistant nanostructured polymers for tires and drive belts, super-hard nanostructured ceramics for drill bits and cutting surfaces, or ultra-fine nanostructured membranes for filters and seals. The second (and, in the long run, probably more disruptive) agenda links nanotechnology to mechanical engineering and robotics. This variety focuses on constructing nano-scale devices for operation in macro-scale environments. Examples of such "nanomechanisms" — or "nanites" — might include ultra-small in-vivo medical devices, miniaturized surveillance systems, or lilliputian mining and manufacturing equipment. Needless to say, these two varieties of nanotechnology blur together at their boundaries: One could, for example, imagine a nano-material filter that employed interconnected nano-mechanism "turnstiles"; or, conversely, one could imagine a nano-mechanism robot that assembled, inspected or repaired nano-material compounds. However, the broad distinction between nano-materials and nano-mechanisms considered as archetypes serves to highlight important "scope conditions" for predictions about how nanotechnology will (or will not) transform the social order.

For many industries and many aspects of social life, nanostructured materials are likely to represent profoundly important technological developments. Indeed, if nanates live up to their early billing, the resulting societal transformations could equal the transformations that attended the development of bronze implements at the end of the stone age or the introduction of nuclear weapons in the 20th century. Given this, it may seem odd to predict that the impacts of nano-materials will be "localized and manageable," as asserted above. My premise, however, is that nano-materials with enhanced performance characteristics do not, in and of themselves, pose unprecedented challenges to social organization. Admittedly, particular new materials may engender the development of particular new products that profoundly transform particular areas of social life — and the ripples from such transformations may spread through society in complex and unpredictable ways. (One need but imagine the implications of a bullet coating that allowed small arms fire to penetrate tank armor, or of a photovoltaic cell that eliminated the need for fossil fuels.) But such transformative potential is hardly unique to nanoscience. Nanotechnology may be more likely to yield these radically new materials, but humanity has developed quite a few other transformative compounds — from glass to gasoline to plastic — without the aid of nano-scale understanding. Thus, although the introduction of new nanates may be revolutionary, it promises to be revolutionary in relatively familiar ways.

Recent years have generated a substantial literature on the industrial impact of technological discontinuities (for a survey, see Tushman and Anderson 1996). In brief, the central finding of this literature is that the introduction of any radically new technology initiates a period of ferment within the affected industrial sectors. During this ferment, businesses, public agencies and other "technology champions" jockey for position, attempting to mobilize various economic, cultural and political forces to frame and tame the discontinuity, and to thereby establish a new "dominant design." Once such a dominant design emerges, however, the affected sectors restabilize — although often in significantly different configurations than before. Presumably, a similar dynamic would attend the introduction of new nano-materials, albeit perhaps at a higher intensity and a broader scale. Of course, this hardly obviates the need for foresightful policy attention: The changes initiated by even a relatively modest technological discontinuity can be quite farreaching, and the framing and taming process can be intensely political, both within the affected industries themselves and at the level of the larger polity. However, the differences between previous technological discontinuities and the discontinuities resulting from nano-materials seem likely to be matters of degree, rather than of kind.

Industries will certainly be transformed, and some of these transformations will be wrenching and risky, but the involvement of nanotechnology will not, in itself, make the transformations any more wrenching and risky than the transformations that we have seen in the past. It is in this sense that the impact of nano-materials will be "localized and manageable": The policy issues, even if large, will arise from the particular performance characteristics of particular products, not from the inherent nature of nanotechnology per se. If so, case-by-case planning would appear to represent an appropriate and, arguably, sufficient response. (As an aside, it may be worth noting that nano-materials could produce unprecedented disruptions if they instigated technological discontinuities in an unusually large number of industries simultaneously. Without entirely ruling out such simultaneity, however, the practical limits on human resources and attention strongly mitigate against this prospect. In all likelihood, applications of nanotechnology will be neither more nor less staggered in their arrival than applications of previous multi-purpose technologies, like semiconductors, synthetic polymers, or wireless telecommunications.)

In contrast to the relatively familiar challenges of nano-materials, nano-mechanisms promise (or threaten) to confront society with policy issues that are as unprecedented as they are profound. By allowing humans to manipulate the world at a previously unattainable scale, nano-mechanisms open a genuinely new frontier, beyond the contemplation of traditional legal and governmental regimes. While, as Richard Feynman put it, "there is plenty of room at the bottom," (Feynman 1961) at the moment there are very few sheriffs at the bottom, to keep that room safe and productive.

As currently envisioned, nano-mechanisms are likely to possess at least three properties that will generate novel safety and governance challenges: invisibility, micro-locomotion, and self-replication.

1. Invisibility is, of course, an inherent property of nano-scale objects, whether natural or artificial. Artificial nanites, however, would be among the first complex constructions intentionally engineered to accomplish human purposes at a microscopic (or sub-microscopic) level, and their introduction into the technological armamentarium would dramatically increase the potential for orchestrated covert activities.
2. Micro-locomotion is less inherent in nanotechnology than is invisibility, since some nano-mechanisms will undoubtedly work best when anchored to substrates. Nonetheless, free ranging nanites have clear advantages for certain tasks, and as such devices are introduced, they will radically challenge traditional understandings of macro-boundaries and barriers. Fences, walls and even human skin are largely open space, at the nano-scale.
3. Self-replication is clearly not an inherent property of nano-mechanisms, and indeed, creating self-replicating nanites may prove to be one of the most difficult technical hurdles of the nanotechnology revolution. Self-replication, however, is essential for the economical production of complex nano-mechanisms in useful quantities, and thus it seems likely that by the time such nano-mechanisms become socially significant, self-replication will in fact be a fairly common attribute. Unfortunately, self-replication also poses profound challenges to human foresight and control, since a population of carelessly designed self-replicating nanites could grow exponentially, without a ready "off switch."

Needless to say, the hazards of invisibility, micro-locomotion and self-replication would be magnified if nanites also possessed a capacity for autonomous operation and selfmodification; however, such higher-order artificial intelligence is not a prerequisite for the envisioned crisis of nano-governance. Even "dumb" nano-mechanisms would require a dramatic rethinking of society's current legal and normative structures. Three governance issues promise to be particularly acute: monitoring, ownership, and control. Moreover, these issues may play out not merely with respect to nanomechanisms themselves, but also with respect to the "nanospace" that such mechanisms render accessible:

1. Monitoring: Invisible, micro-locomoting, self-replicating nanites will severely test society's established assumptions about what can and should be monitored, and by whom. On the one hand, the lay public may find itself inundated (or even interpenetrated) by complex devices whose presence, provenance and purpose remain undetectable without sophisticated technical assistance. On the other hand, those actors that can produce or purchase nanites will find themselves able to monitor their worlds (including their

social worlds) in more profound and surreptitious ways than ever before. Contrary to traditional assumptions, ordinary people will no longer be able to observe all the socially relevant activities in their own surroundings; yet public authorities will be hard pressed to provide policing assistance without further endangering individual privacy. In the absence of trustworthy institutions to regulate this newly-opened invisible frontier, the potential for rampant abuse will be matched only by the potential for rampant paranoia.

2. Ownership: Nano-mechanisms will also severely test society's assumptions about what can and should be owned, and by whom. Self-replication raises difficult questions of whether property rights persist from one generation of mechanisms to the next — not to mention difficult practical problems of "branding" proprietary nanites and policing "nanite rustling." Equally importantly, micro-locomotion raises difficult questions of who owns the nanospace through which nanites pass. Controversies akin to historical debates over airspace and rights-of-way may emerge over the nanospaces within macroscopically "solid" objects. If your nanites get sucked into my air conditioning system, are you trespassing or am I kidnapping? Can public health authorities claim an easement for the passage of biomedical nano-sensors through my gut? If nano-robots in my tap water "harvest" a microscopic quantity of copper from my plumbing, is their owner guilty of breaking and entering?
3. Finally, and perhaps most significantly, nano-mechanisms will test established assumptions about responsibility and control. Traditionally, legal liability for mechanical devices can outlast ownership, but can be attenuated by unforeseeable circumstances or by the intervention of a third party's will. The viability of these principles, however, becomes unclear in a world of self-replicating nano-mechanisms that operate invisibly in an only partially understood micro-environment. One might imagine governing the control of nano-mechanisms by analogy to defective products, or to toxic emissions, or to speeding vehicles, or to straying livestock — but the implications of these analogies are not entirely congruent with one another. Moreover, a parallel set of complexities and contradictions surround the control of nanospace: Who is responsible for policing the boundaries between "open range" and "enclosed territory" at the nanoscale? Can negligently maintained nanospace pose an "attractive nuisance" to passing nanites? Does a state maintain territorial rights to the nanospace within the bodies of citizens traveling abroad?

Although the distinction between nano-materials and nano-mechanisms is to some extent an artificial one, it highlights an important dimension of differentiation in the impact of nanotechnology: Nanostructured materials may pose serious practical and ethical challenges for particular policy domains, but these challenges will arise at a familiar macro scale, for which we have numerous rules, institutions and historical precedents. In contrast, nano-engineered mechanisms will force us to reformulate our rules and institutions to govern an unfamiliar setting with which we have no prior experience. By allowing human will to operate in the previously unreachable interstices of macro-reality, the nanotechnology revolution promises to open vast new frontiers for exploration, exploitation and settlement. Taming those frontiers, however, will require us not only to control physical matter, but also to control ourselves.

In the past my thoughts and ideas about nanotechnology have flowed easily, as they were prompted by the excitement and prospects of a coming new wave of scientific challenges and opportunities. Putting together a statement on the societal implications of nanotechnology has proven more difficult. During the past six months I have asked a number of people what they thought nanotechnology meant, and what they thought the societal implications might be. While most responses were positive, they varied depending on the degree of technical background. Most scientists and engineers saw nanotechnology as an encompassing technology that would broadly impact on microelectronics, data storage and the computer area, energy conversion and chemical processing, defense and national security, biotechnology (including agriculture), and the biomedical areas of therapeutics and diagnostics. While some scientists and engineers felt that nanotechnology was not quite as "new" as perceived and presented by the popular media, they did agree that a new level of much higher activity and broader applications was definitely happening. For the most part, this group felt that the overall implications of nanotechnology would be very good. In fact, some felt it might even be vital that we progress as rapidly as possible into this new area. The feeling was that nanotechnology (as well as other related technologies) may provide solutions to some of the more intractable problems related to eliminating disease and famine in developing countries, and helping to improve the economy and general living standard for the rest of the world. In general, scientists and engineers that I talked with tended to feel nanotechnology would carry risks similar to those seen in the development of other previous technology revolutions, e.g., microelectronics. The feeling was that, while nanotechnology could certainly be used for some insidious purposes, the overall benefits would far outweigh the risks. If there was concern about nanotechnology, it centered more on the need to be careful not to oversell or make inflated promises about nanotechnology, as this could lead to loss of credibility and confidence by the general public. Responses from non-technical professional people were different from those of the scientists and engineers. In general, this group was not as well aware of the potentially more far-reaching applications of nanotechnology (and this would not be unexpected).

This group tended to equate nanotechnology primarily with the further miniaturization of microelectronic devices and very advanced computers. Some from the non-technical group related to nano-type medical devices, which might for example be put into the blood stream of humans to treat cancer or other diseases. While this group may not really understand nanotechnology, they generally thought that it was both very interesting and important, and that it might provide many future benefits. Some from both non-technical and technical groups do have an underlying uneasiness with how quickly technology in general is advancing. This uneasiness could most certainly reflect on nanotechnology.

Some concerns about nanotechnology (and technology in general) are not just related to deliberate or insidious misuse of these technologies, but also include problems which could result from carelessness or unexpected adverse effects which appear later in implementing the technology. By way of previous examples, the dangers from nuclear power plant breakdowns, the pollution from chemical processing and toxic reactions to certain therapeutic agents. While these types of concerns are justified, most people still believe that technology will of course continue to advance in the future and will most certainly provide benefits for all of mankind.

This is true even more so in a field where two different systems of science come together. Natural scientists and technologists have to understand elementary notions of social science, and social scientists in turn have to learn about the scientific and technological object of their study to do a meaningful analysis. The basic vocabularies have to be learned yet. The objective of this paper is to provide a European point of view on the social implications of NST. Past and ongoing efforts in the social sciences that are related to NST are introduced. Part I of this paper will give an overview of mainly European studies in this area. Over the past few years, a modest amount of activity of social scientific research has been carried out in Europe that specifically addressed economic and social issues of nanotechnology. A review of these activities might provide a useful platform for further discussions. Part II will give an illustration of a quantitative social scientific approach to NST. This is done to clarify where social sciences can contribute and where inputs from actors in the field are needed. Finally a number of conclusions are drawn and some suggestions are made with respect to where future social scientific research on NST might be of interest.

Economic and Social Aspects of Nanotechnology: One should make a distinction between activities in the social science arena that are directed at describing the emergence of nanoscience and technology as it happens and those activities that deal with issues that might become relevant once nanotechnology has entered a more mature stage in its development.

Studies on Current NST Activities or Near-Term applications: Often and especially in a European context, nanotechnology or nanoscience is discussed in terms of potential paths towards commercialization, niche markets for already or soonto- be available products. This is also the major focus of ESANT, the European working group for

Economic and Social Aspects of Nanotechnology, a small community of researchers and practitioners, mostly with a technological background. The group has published a report (http://www.nano.org.uk/ESANT99.htm). In their report, potential avenues of near-term commercial development of nanotechnology are outlined.

Other social science research focuses on how governments support nano-scientific research and the implications this has for the development of nanotechnology. One example is the work commissioned by the Finnish National Technology Agency (Meyer 2000b). Nanotechnology is addressed in many different ways. The different perspectives observed at the country-level seem to coincide with the strengths of the respective national innovation system. There is no common definition as to what nanotechnology comprises. It appears that specific scientific and technological communities have reached a common basis as to what "nano" means with respect to their particular area. The current state of development allows government actors to play a substantial role in developing a novel technological area. In some European countries, governments rely to a considerable extent on technology foresight studies in their long-term science and technology policy planning. These exercises provide governments with estimates about future technological developments and the implications they might have for society. Some of these studies also addressed NST. We shall have a closer look at them in the following section.

Mid- and Long-Term Issues: Foresight studies attempt to depict an image of a possible future using a variety of techniques. An important foresight method is the Delphi survey. A Delphi survey basically is a tool to create consensus and detect areas of conflicting expert opinions. In a first round, experts are confronted with a number of topics they have to evaluate with respect to time of realization, implication on wealth creation, quality of life, and similar issues. In another round the results of the previous round are introduced to the experts who then have a chance to re-evaluate the topic. An overview of the nanotechnology-related topics covered in the 1995 German Mini-Delphi survey. Kuusi and Meyer try to distill five major areas of developments ("*leitbilds*") drawing on this Delphi exercise and information from other technology studies (Kuusi 1999). *Leitbild* is a loan word from German, meaning a guiding image. Within the context of the social studies of science and technology, it can denote developments in unfolding areas that might lead to the emergence of a novel paradigm or technological trajectory. Its major function is to provide a platform integrating actors from different areas by providing a common goal, which can be an envisaged technique or product.

1. Nano-resolution methods of analysis. One *leitbild* is that of nano-resolution analytical methods (topics 20, 22). The aim here is to further improve existing tools by adding new functions to analysis tools. Here, realized techniques are further generalized into promising tools. The development of analytical tools and techniques is closely related to progress in other areas. Often nano-resolution tools are identified with nano-resolution optical microscopy since these methods combine the possibility of

measurement with that of manipulation, which makes them more versatile as tools for nanotechnology. A number of scanning probe microscopes have been developed and some of these methods do not require complicated sample preparation; e.g., the ability to work on *in vivo* substrates and determine structurefunction relationships is the main reason why the AFM, the atomic force microscope, is so popular amongst biologists. It is said that even (nearly) two decades after its invention, the implication of the scanning-tunneling microscope (STM) and its follow-ups is still growing. In conjunction with continuous technical further development of this method, researchers are discovering new phenomena in the fields of physics, chemistry, and biology. At the same time the STM methods are used as a nano-tool rather than a nanoprobe. The idea is to modify surfaces and tailor their structures on the nanoscale, down to the manipulation of individual atoms (Frenken 1998). This case has shown that realized techniques — such as AFM and STM — can provide a basis for developing more sophisticated "promising" techniques that will go beyond mere measurement. Ultimately, they might facilitate large-scale manipulation at the nanometer level.

2. Nano-materials. Another *leitbild* ranges around the notion of nanostructured materials. Nano-materials can be seen as structures that have particular properties owing to their size. The *leitbild* in this context is to manufacture a nano-material in a way that allows predetermination of its properties. This generalization pattern describes the transition from realized techniques to promising targets. Paired with a better scientific understanding of the subject matter, a variety of already realized techniques allow us to develop rather concrete ideas of improved materials. The idea here is to take advantage of nanoscale characteristics of structures and substances to create new materials with enhanced properties, such as those common to polymers, composites, or other materials (topics 16, 17, 19). The aim is not necessarily direct control of individual atoms; bulk operations suffice to exploit the nanoscale properties. One example to illustrate the idea of bulk-processing nano-materials is that of colloidal dispersions (Philipse 1998). Colloidal science deals with the physics and chemistry of finely dispersed matter. Colloids are generally understood as particles or other objects with at least one dimension roughly in the sub-micron range. Nano-particles then are viewed as colloids smaller than 100 nanometer. It is pointed out that colloid science has a long tradition involving nano-particles and that not all that is nano is necessarily new (Philipse 1998). In this sense, colloids encompass gold colloids, colloidal silica, and aluminum oxide powders. Due to their small dimensions, colloids exhibit Brownian Motion. Owing to their large surface area, surface forces and repulsions determine the interaction between colloidal particles in the liquid phase. The balance between these forces depends critically on the details of the particle surface and the liquid composition. Colloids aggregate easily to form large flocs, networks or gels. The control and understanding of these aggregation processes is the major challenge of colloidal nanoscience. Suspensions of colloids, or dispersions, are, for example, milk, blood, cosmetics, such as toothpaste, or ink. Examples for inorganic colloids are clay particles, iron oxides on computer disks and in magnetic fluids, pigments in paints, or powders

for technical ceramics. Colloidal systems are studied from a (bulk) chemical synthesis perspective. Such research integrates the study of physical properties of dispersions and gelation phenomena, obtaining important input from computer simulations and statistical mechanics. It is said that this wide scope of colloidal science is unavoidable, since finely dispersed matter can be encountered in many disciplines and applications.

The nano-materials *leitbild* provides a focus point for integrating this and other bulk chemistry topics with related activities.

3. Ultra-thin layers. A number of topics also revolve around thin layers (topics 14, 18). Here, efforts appear to be directed at characterizing these structures. Thin-film technologies constitute a considerably well developed field. An exact ultra-fine production of thin films is a necessary condition for the subsequent characterization. Designing ultra-thin layers can be associated with a small number of technical aims, such as atomically exact delineations of layers, quantumized potential distribution, defined pore distribution in layers, ultra-thin separation and protection layers, improved layer function by way of multilayer structuring. These technical targets are related to a greater number of applications. Examples are information storage layers, films with quantum effects, optical layers, multilayer piles for quantum/semiconductor laser and x-ray optical compounds, displays, sensor layers, tribologic films, biocompatible films, photovoltaic films, membrane films, and chemically active surfaces (Bachmann 1998). One might find that activities within the context of this *leitbild* follow a generalization pattern where promising targets are used to develop promising techniques. Realized techniques permit already sufficiently exact operations at the nano level to convey an idea of future products. The idea of future products, in turn, requires even more exact and precise tools.

4. Bottom-up techniques. While the preceding *leitbild* is still to be positioned in the field of top-down nanotechnology, the subsequent *leitbild* addresses the nanoscale from a different, "bottom-up" perspective. This perspective attempts to simulate nature to develop materials with novel properties by way of self-organization. These approaches are often dubbed "biomimetics" (topics 19, B). In this case, an integration of various techniques is necessary. It has been shown that one can create structures by way of self-organization in a biomimetic process. However, our technological means are inadequate to fully utilize the potential this *leitbild* offers us. Being aware of the general feasibility — thanks to already realized artifacts — we can make reasonable assumptions about the requirements of the techniques necessary to pursue this path of development further.

5. Direct atomic control. Another *leitbild* focuses on direct control of atoms, to rearrange them into new atomic structures (topic 15). This could lead to novel materials. The difference between the materials approach mentioned above and this one is related to how one controls the process (bulk reactions *vs*. atomic control). Atomic control is also strongly related to the idea of atoms being effectively used as carriers of certain functions, such as data storage, etc.

Societal Implications of Nanotechnology: The workshop's efforts are chiefly directed at identifying areas where mature nanotechnology can influence society. In Europe, the ESANT working group has undertaken the effort to outline possible avenues of societal implications of nanotechnology. The idea was to "foresee some beneficial aspects of nanotechnology for society in general" that reach beyond mere economic benefits, a topic a number of studies focus on currently.

Even though the ESANT study stresses the beneficial role nanotechnology is expected to play in economic and societal development, the group felt it is necessary to point to a possible detrimental effects and risks that may be associated with nanotechnology.

To some extent, parallels are drawn with biotechnology, even though it is pointed out that that broad-brushed, general comparisons won't do justice to the technological field. Nanotechnology may lead to artifacts that incorporate genetic material or have genetic modification or repair as an objective and in this context, risks associated with genetic engineering may be of relevance also in terms of nanotechnology. Other areas that may involve some risk that should be taken into consideration are implications of the physical size of artifacts that are based on nanotechnology. A further group of risks is related to the self-assembling or self-replicating nature of foreseen nanotechnological processes. Finally, nano-composites were viewed as potentially increasing environmental problems. The group points to the generic character of nanotechnology that "places it in the same position as only two other technologies in recent decades, genetics and information technology. Of the latter two only genetics has received any extensive examination in terms of desirable applications and risks; information technology has escaped this kind of critical examination completely. Since nanotechnology embraces the entire set of genetics and information technology it is of paramount importance that its risks be speculated about in advance of desirable implementations, but not in an uninformed way" (ESANT 1999). The role of the media and a better understanding of this process was viewed as a critical factor. Furthermore, it was pointed out that *"methods of assessing risk, particularly those relating to toxicity, may themselves be inadequate to the new situations posed by applications of nanotechnology in many different fields."* If this was indeed to be the case, one should be aware of the time that could elapse *"before this [inadequacy] was recognized, allowing public perceptions of the risks involved to influence policy formulation in ways that would be difficult to reverse"* (ESANT 1999).

Therefore, it is important to question the judgements made and assess the process that has generated them in terms of its validity. This goes beyond specific techniques. Depending on the process and various other aspects, the "quality" of the judgements can vary substantially. One might call this "quality" of statements their *"epistemic value"* (Kuusi 1999b). Generalizations, or projections of certain technological developments, as well as judgements of their societal implications should be exposed to a critical discussion of their epistemic value. There is some, yet unpublished work going on in

Europe that addresses these aspects and attempts to develop a methodology to assess the epistemic value. The work is carried out at a general level, but can also be applied to the context of NST. The gist of the concept is summarized in an equation. From the point of view of some actor k, one can describe the epistemic utility value DUk as an evaluation of a technological generalization, or societal implications, as follows (Kuusi 1999b): DUk = Uk (I1, F1, V1 k, R1 k, Lk)—Uk (I0, F0, V0 k, R0 k, Lk)) The idea in the formula is that the epistemic utility of an evaluation for an actor depends on the value of five components after (I1, F1, V1 k, R1 k, Lk) and before (I0, F0, V0 k, R0 k, Lk) the presentation of the realized technology generalization/argument.

It is typical that the anticipated implications (I) of a topic considered in the bootlegging stage are not identical with the realized implications in the bandwagon stage. If a realized generalization/argument changes the assumed implications, I1 and I0 differ. An argument might also change the suggestion concerning the feasibility (F) of the used techniques, which means that F0 and F1 differ. The formula supposes that all relevant actors k understand the suggestions concerning implications and used techniques in the same way. Based on this assumption F and I are not related to specific actors k. The differences of actor opinions are focused on the validity (V) and on the relevancy (R) of suggestions. It is reasonable to assume that in the bootlegging stage the evaluations concerning the validity differ considerable both inside the group of protagonists and especially between protagonists and outsiders. The proponents of the old paradigm do not believe that the proposed techniques could produce the proposed implications, or they believe that beside proposed implications the techniques produce other questionable implications. Opinions might also differ considerably concerning the relevancy of produced implications. R1 k and R0 k are the relevancy of the topic or the technology generalization to the actor k after and before the presentation of the realized generalization/argument.

If we consider the suggestions concerning future action, the reasonable realization of a topic depends on the other choices open to k. Lk describe these other choices. The model assumes that the realized technology generalization/argument has no implication on the epistemic value of competing choices based on other technologies The research presented here is a theoretical model. It might be difficult to calculate such an "epistemic value" in practice. However, the underlying idea of the model is of great relevance. Contributors to studies of societal implications of NST must be aware of the varying validity and relevance of their assumptions and should address this issue in some structured way.

This study will introduce a different way as to how one can approach nanoscience and technology from a social scientific viewpoint. This approach is widely quantitative, drawing on bibliometric and patent data. The study that is to be introduced here can illustrate the possibilities of social science research on current NST. It will also show the limitations of such approaches and identify areas for dialog between natural scientists and technologists on the one hand and social scientists on the other. Social science might help scientists and technologists put their work into a broader context. The quantitative approach we are going to introduce here will not be able to tell us where NST will be in 25 years from now, but data like ours can be used to identify scientific and technological trends over a period of up to ten years. The data can convey an impression as to where nanotechnology originated and indicate the general direction in which the area moves. The bibliometric analysis we are going to present can also be used to classify the field itself and indicate to what extent work related to nanotechnology takes place in a disciplinary context. It can also help track the connection of science to patents.

However, one must be aware of the limitations of such approaches. They include in particular the key words approach used in data retrieval and the delineation of the subfields of NST. Publication and patent data was retrieved using a key word approach.

These keywords always include some subjectivity. We used terms that were applied originally by bibliometricians, not experts in the field. It should also be mentioned that the publication and patent subfields were delineated on the basis of established classification systems. These systems are structured after disciplines and technological sectors. This might not always be the most appropriate form of organizing a field-specific database.

In order to analyze knowledge generation in nanotechnology, we set up three different databases:

- A bibliometric database of all scientific research papers related to nanotechnology
- A database of all U.S. patents in that technological area
- A database that connects both data sets by identifying the references citing from nano-patents to nano-papers

While the bibliometric database allows us to have a look at how transdisciplinary current activities in nanotechnology-related sciences are, the second (patent) database offers a perspective at what has been done in terms of technological development so far. The third database will be instrumental in determining the extent to which scientific knowledge generation is related to its specific "application context," namely nanotechnology. We use the number of patent citations as an approximate measure for relevant connections.

1. Bibliometric Database. The bibliometric database contains 5,000+ scientific research papers on nano-related subjects. Following Braun et al., we retrieved all publication titles of journal articles in the Science Citation Index and patents downloaded from the USPTO Internet database that included the term "nano" as such or as a prefix.

Subsequently, irrelevant records that contained only terms like nanogram, nanosecond were deleted. The Appendix contains a list of search terms. The publication data is analyzed at various levels, one of which is distribution by major and sub-disciplines.

Looking at the percentage distribution of nano-papers one can establish some trends. We calculated a linear slope coefficient based on the annual shares of major fields as well as subfields. While natural and multidisciplinary sciences have gotten stronger, the major fields of engineering and materials and life sciences seem to have lost importance. Papers in natural science have a slope coefficient of 3.2. The only other growing field is multidisciplinary sciences with 0.4. Engineering and materials as well as life sciences lost ground dramatically. Their shares dropped by more than a half and three fourths, respectively, leading to coefficients of -2.3 and -1.5.

A look at the subfield results gives a more detailed view. There we see that the growth of nano-papers with a natural science classification is not uniform. Thus the slope of chemical nano-publications is twice as steep as the one of physical nanopapers. Publications is due to the subfield of interdisciplinary natural-engineering and materials science. While the shares of other multidisciplinary sciences are decreasing, this subfield could increase its share by more than a third. Its slope coefficient equals 1.3. An interesting observation in this connection is the decrease in share of publications in materials science (-1.9). This more than what happened to the other multidisciplinary fields seems to have contributed to the growth of that particular subfield.

2. Patent Database. The patent database comprises approximately 2,600 patents, mainly in instrumentation, electronics and electrical engineering, and chemicals/ pharmaceuticals. The development of the database over time and the distribution of patenting after technological fields. One can recognize a clear focus on instruments, electronics, and chemicals/ pharmaceuticals. An interesting trend one can make out is the strengthening of the chemicals/pharmaceutical sector. An update of the database (done in 1999) would show if this trend prevails. As the USPTO will make available patent application data in the near future, the analyses should be significantly improved since the time lag will be minimized. It might be useful in terms of the societal implication efforts to have a closer look at a number of areas, trying to identify paths of technological development and subsequent industrial applications. Based on these trends, one could develop scenarios of potential societal implications these applications might have.

As these sectors are generally attributed a certain proximity to the science-base, one should expect a substantial overlap between nanoscience and nanotechnology, indicated by a considerable number of relevant nano-research papers being cited by nano-patents.

3. Patent Citation Database. The patent citation analysis confronted us with somewhat surprising, counter-intuitive results. Linking scientific and technological databases can go beyond evaluating the intensity of exchange processes in an emerging field. It can also help illustrating and understanding if and how new developments in science and technology might find their way into society. We can follow up how technological sectors are linked to certain domains of science in 3 One illustration of this is, for instance, a major EU-funded conference on 'Nanoscience For Nanotechnology'; the title implicitly assumes that nanotechnology builds critically upon nanoscience.
4. It should be noted that the matches we found are patent citations listed in the 'other references' section on the front-page of the patent, referring to scientific papers in our nano-publication database only. This methodology corresponds to current practice amongst patent bibliometricians who track front-page citations of non-patent literature to study the science/technology linkage. The following section on the nature of patent citations will clarify the importance of this practice.

NST. This again can be specified for various industries and organizational channels. This again may allow actors to formulate educated guesses about future developments.

If one looks at nano-patent citations by industrial sectors and SPRU-main classes, the instruments sector appears to be the sector that relies on nanoscience the most (with 49 citations). Instruments are followed by electrical machinery (40 references), electronics (35), pharmacy (29), and chemistry (21). Other sectors refer to scientific domains less than 20 times. The most referenced scientific domain is natural sciences (145 citations), followed by multidisciplinary sciences (87). An overview of the distribution of the patent citations. Not only is the overall interaction between nanoscience and nanotechnology relatively weak with 275 5 Given the dominance of natural sciences it might be interesting to have a closer look at the level of subclasses. The subfields that were cited ten times or more are: physical sciences (with 108 citations), multidisciplinary sciences (40), interdisciplinary naturalengineering and materials (36), chemical sciences (34), materials science (19), medical sciences (14), and inter-disciplinary lifenatural (10). The data indicate clear differences in the linkage patterns of the industries investigated. While instruments, electric machinery, and electronics have a strong focus on physical sciences (with 19 out of 49 nano-patent citations, 23/40, and 24/35, respectively), the pharmaceuticals and chemicals industries have different orientations.

For instance, with ten out of its 29 citations, the pharmaceuticals industry is linked to the medical sciences much stronger than any other industry. Similarly, it is connected to inter-disciplinary life-natural by 8 patent citations. In terms of nanoscience and technology, the chemicals industry seems to put emphasis on inter-disciplinary natural-engineering and materials and multidisciplinary sciences to a relatively greater extent than instruments, electric machinery and electronics industries. Various combinations of these and other sets of indicators confirm the results presented here. For instance, the distribution of patent citations as measured by Dewey descriptors and IPC classification at the 4-digit level present similar results. Physics/engineering with 66 citations is the major 'contributing' scientific domain. The strongest connection has been established with subclass H01L (20 citations). Other links are with H01J (6 citations) and G03F (4 citations).

The second biggest contributing 'science' is the field of comprehensive works and general sciences with 37 references, which are much more evenly distributed. With 6 citations, H01J tops the list of technological domains linked to this field of science. Runners-up are H01L and C01B. The scientific field 'physics' holds rank #3 with 25 patent citations. However, there is no technological field that more than 3 citations are related to. The Dewey-domain 'engineering/chemical engineering' is the fourth biggest attractor of nanopatent citations, together with 'pharmacology and pharmacy' (both 15 references). These two areas, however, relate to nanotechnologies in entirely different manners. While (chemical) engineering is referenced just by B32B four times and linked to eight other technological domains with one or two patent citations, 'pharmacy and pharmacology' are clearly linked to subclass A61K (with 12 out of 15 patent citations). There are three further linkages established by individual patent citations (A61F, C07F, B01J; one citation each). The only other 'stronger' link is the connection between the scientific field of instruments and the IPC subclass H01J with 4 references. The following tables present a number of examples arranged by field and organizational category. Thereby we can gain a perspective on how patent citations describe the science/technology interface in organizational terms for a specific subfield.

The case of nano-patent citations in the field of *"other machinery"* illustrates a reference pattern in which nanoscience-citing university nano-patents substantially rely on scientific nano-publications authored in the university system or non-industrial research centers. Similarly, it also shows that respective patents assigned to multinationals chiefly refer to work from their own organizational category. The case of *metal products* points to the universities as the main producer of nanoscientific information relevant to nano-patents. However, it also demonstrates the universities are the most citing assignee organization in this area. This might raise questions as to how "industrial" academic patents are. The case of *electronics* illustrates the importance of universities as science producers. However, universityheld patents cite university-generated research much more frequently than others.

Industry, in the form of multinational corporations, appears to be an important producer of scientific knowledge and also a major absorber. The case of *nano-instruments* indicates the importance of intermediary organizations in terms of knowledge diffusion. Industrial associations and similar organizations seem to be the second largest absorbers of knowledge in the field. Again, it can be pointed out that university-papers are most frequently cited in university patents. These examples illustrate insights that can be gained from the application of patent citation analysis at the combined sectoral/ organizational levels. We could demonstrate that the NST cluster is very heterogeneous and its different subfields show rather varied patterns of knowledge transformation, diffusion and absorption.

What can we learn from above analysis? First, we have realized that science related to the nanoscale takes place mainly in a disciplinary context, and, therefore, we are far away from a unified nanoscience. However, we were able to point to a relative increase of natural scientific and multidisciplinary publications and a relative decrease of materials sciences. The patent database we set up illustrated the broad range of technological artifacts and processes where the nanoscale matters. An analysis of the citation linkage between nano-patents and nanoscience indicated that there is just a weak direct connection between technology and science at the nanoscale. Moreover, an investigation of patent citation has shown that many of the patents that do cite nano-papers are assigned to universities. This could indicate a process of knowledge transformation from science to technology within the academic sector rather than a knowledge transfer process from academe to industry. Further analysis also pointed to sectoral differences as far as who cited whom. There were different organizational patterns to be observed. The data should have illustrated that NST will develop various aspects in many, different ways. And the citation analyses also should have shown that each of the areas NST can affect follows different transformation and diffusion patterns. Studies on the implications of NST for society will need to pay attention to this complexity. The tool introduced is a good method to map activities in NST. At more advanced stages it can also be used to track more specific elements. Using such a social scientific methodology on a NST can also be instrumental in facilitating a basis for a common language. The results of this study could create a discussion about delineation and character of NST and their various sub-fields. This discussion will familiarize natural scientists and engineers with social scientific thinking while it forces social scientists to deal with scientific and technological aspects of the heterogeneous and complex field of nanoscience and technology. This mutual understanding is of fundamental importance if one is to study societal implications of nanotechnology. Social scientists need to gain an understanding of where the new technology is coming from to be able to discuss potential societal implications of nanotechnology in a meaningful manner. This approach also would counteract a simplified understanding of mature nanotechnology as the realization of the assembler visions of the molecular manufacturing pioneers. The major point this paper wants to make is that societal implication studies of (mature) nanotechnology should not start with unsubstantiated visions of the future but begin with an effort to understand NST as it emerges. The social implications program of workshops provides the chance to start a dialog between scientific and industrial experts and social scientists in the area. Social scientific research like the contributions presented could form a platform to build on. Some of the European studies introduced in Part I underline the point made at the workshop that in particular areas it makes sense to analyze past experiences with other technologies to prepare for future experiences with nanotechnology. Biotechnology is an apparent candidate for this undertaking. However, one must be aware of the risks of unspecific comparisons. Finally, another area that needs to be addressed is the process of evaluating and assessing potential societal implications itself. European studies on NST pointed to the need for measures that can control reliability and validity problems that come along with implication research on future technologies. It is important to find a solution for this issue or at least a way of communicating its implications if one wants to ensure that the implication information generated will be used in an appropriate manner.

Radically new technology inspires lyrical utopianism and melancholy catastrophism. What is new at the turn of the twenty-first century is the note of alarm among the leaders of change. Eric Drexler and Bill Joy have focused on the self-replicating potential of nanotechnology processes. From the outset, Drexler has acknowledged the possibility of rampant synthetic organisms that could displace real ones (the "gray goo problem,") and Joy has even speculated that quasi-human robotic systems constructed with nanotechnology could in effect enslave our species. But there is old-fashioned optimism, too. Enthusiasts foresee agricultural bounty, a paradise of health and longevity, mental and physical enhancement, and a wonderland of novel consumer goods. The web site of a forthcoming lay publication, *NanoTechnology Magazine*, promises "friendly energy" and "positive impact agriculture," the end of animal experimentation, and the neutralization of all chemical and even radioactive waste (http://nanozine.com). The history of technology can not reconcile these visions. But it can help prepare us for the surprises that have always been the result of human ingenuity. We can expect five things: (1) The experts will be seriously wrong about at least some important things. (2) Long-term, cumulative problems will be a greater problem than the perils of catastrophe. (3) Organizing and supervising nanotechnology will create dilemmas. 4) Successes may be as costly as failures. (5) We probably have not imagined the greatest benefits of nanotechnology, either because they seem too technologically modest or because they may result from improbable chains of events.

The most gifted scientists and engineers have a mixed record as long-term forecasters. Lord Kelvin, outstanding as an inventor as well as physicist, is now also known for his prediction that heavier-than-air flight would remain impossible. Irving Fisher, equally eminent a generation later as an economist and entrepreneur, declared in 1929 that the stock market appeared to have reached a permanently high plateau. In a 1955 book, *The Fabulous Future*, published by *Fortune* magazine, John von Neumann predicted that by 1980 improvements in reactor technology would make nuclear energy so economical that it "may be free – just like the unmetered air — with coal and oil used mainly as raw materials for organic chemical synthesis...." Following earlier scientists, he did recognize the human influence on what later became known as global warming, but he also foresaw advances in atmospheric science leading to deliberate climate modification "on a scale difficult to imagine at present," including the possibility "of a new ice age or of a new tropical . . . age in order to please everybody." Yet von Neumann was equally far too modest in his expectations for the electronic computer. He imagined its future as ever more powerful central control of economic planning and industrial processing. His Cold War vision for the device he had helped to create had no place for the miniaturization and decentralization of electronics (von Neumann 1956). Of course scientists and engineers imagine the forms their ideas will take and the consequences they will have. The operations of an agency like the National Science Foundation depend on projections of benefit to humanity as well as abstract knowledge. But just as many writers are surprised by the uses to which their ideas are put by readers, innovators must be prepared to find that reality has other plans for their work.

Unlike von Neumann, we know the outcome of the Cold War and how it surprised liberal and conservative forecasters alike. Nuclear energy is neither as threatening nor as immediately promising as expected. But our experience in the 45 years since von Neumann's essay suggests some lessons. In our imagination of technological consequences, we think first of disasters. But for the past century, our concerns have been misplaced. Technology has instead tended to replace catastrophic problems with chronic ones. In place of the threat of an apocalyptic thermonuclear exchange that would be over in days we face the dilemma of dealing with lethal nuclear wastes, some of which will last for hundreds of thousands of years. Freon coolant protected people from oncecommon refrigerator explosions, but at the price of slowly depleting the earth's ozone layer. Low back pain and cumulative trauma disorders like carpal tunnel syndrome now are more serious industrial hazards than loss of life and limb. Where effective treatment is available, AIDS has become not the new plague that so many feared in the 1980s but a long-term, controllable condition. Post-World War II reinforced construction techniques have generally been safe but not necessarily stable. Architects and engineers failed to predict the many interactions of materials, moisture, and weather. Rehabilitating these once advanced buildings has become a complex and costly specialty. It takes time but it, too, is manageable. We defuse problems by diffusing them. There is no guarantee that the future will be like the past. But if the trend of recent decades continues, the hazards of nanotechnology will not be apocalyptic "gray goo, "uncontrollable self-replicating substances or organisms. Chemists and physicists in nanotechnology research argue persuasively that self-assembling "nanobots," drawing resources from the environment as living creatures do, will never be possible.

Nanotechnology, alone or with other innovations, nevertheless might indirectly promote some existing pathogen or parasite. Early in the twentieth century many foresaw the uses of aircraft for war; far fewer realized how effectively mass civilian air transportation and vastly expanded, containerized seaborne commerce would inadvertently help new disease agents cross ocean barriers, including HIV and the West Nile virus. Likewise the spread of mad cow disease (BSE) as human Creutzfeldt-Jakob Disease (CJD) in England arose from the interaction of innovations: changes both in animal feeds and in methods of extracting meat particles from carcasses. The Legionnaire's Disease bacillus has always been common in natural waters without causing epidemics; modern heating, ventilating, and air conditioning systems gave it rare, ideal conditions for reproduction and transmission. Risk analysis might show many technologies to be safe in isolation, yet the range of possible interactions remains beyond it.

Present testing techniques are not consistently able to identify slow, gradual change. Yet environmental damage is most likely to occur in some apparently innocent indirect result of an apparently benign process, just as we are now discovering the problems created by decades of the distinctive chemical wastes of electronics manufacture. One of the greatest challenges of any new technology is determining which new potential problems need to be identified and measured. Diesel engines, for example, produce relatively few pollutants associated with conventional internal combustion engines, but have unique emissions problems of their own. It took time to develop new, appropriate tests for them. R. Flagan and D.S. Ginley, in their chapter on "Nanoscale Processes in the Environment" in the NSTC report (NSTC 1999) observe that we still barely understand how nanoparticles and nanostructured materials affect living organisms and other aspects of the environment. Nanoparticles could set off subtle changes in plant or animal tissues that could cascade into extensive biological change, just as DDT did in the postwar years after it was hailed as an environmental breakthrough. And a series by Tom Horton and Heather Dewar in the *Baltimore Sun* during the week of 24 September 2000 reveals how the most important chemical innovation of the early twentieth century, the chemical extraction of nitrogen from the atmosphere, helped end famine and increased the world's population by an estimated two billion people — but has also slowly and indirectly choked the life of rivers and coastal waters by depleting oxygen. The leading authority on global nitrogen, the Canadian environmental scientist Vaclav Smil, has called the Haber-Bosch process that makes modern fertilizers possible the most important invention of the twentieth century.

Slow, unforeseeable processes may also affect engineering applications of nanotechnology. While smart concrete may be designed to counteract the corrosion that has beset the conventional product, how can it be formulated to anticipate environmental changes brought about during its lifetime by other new substances, including other new products of nanotechnology itself? The very diversity and excitement of nanotechnology research may paradoxically be a hidden weakness. How can we anticipate real-world interactions when we are going to be modifying reality on so many fronts, and when data can take time to interpret? A hundred years after the first observations of global warming, there is still debate over how much of it is due to human causes. And how can we expect to reverse or even curb chronic problems when they take so long to become serious that we have become dependent on the products that have caused them? Nanotechnology may well help us, directly or indirectly, address nitrogen and other environmental issues, but it is equally likely to start changing things before we learn which changes to monitor.

The social issues of nanotechnology will be at least as complex as its consequences for the environment and design. In his book *Normal Accidents*, the sociologist Charles Perrow has argued that some technologies are both nonlinear and tightly coupled. That is, they can feed on themselves *and* a single mistake can be readily transmitted through the system (Perrow 1984). Nanotechnology may reduce these forms of risk significantly, especially by promoting energy efficiency and making it possible to reduce or eliminate nuclear power production. But it is also possible that some otherwise extremely desirable nanotechnology processes may meet Perrow's definition. They may require the kind of strict controls that we now associate with the nuclear industry — controls not only on operations and physical access but also on the sharing of knowledge. To von Neumann's generation these appeared the inevitable costs of a national security state. In the early twenty-first century, as the problems of America's national laboratories show, this system of command is harder to sustain. Knowledge has become far less viscous. Laptop computers and Web connections are sabotaging decades of carefully defined security procedures, yet we are warned that stringent new security could wreck the morale of the producers of classified knowledge. Innovations flow readily across national borders because the sheer volume of world trade and communication overwhelms controls. The programmers in recent headlines have been Finnish, Swedish, and Belgian; tomorrow's may be from any part of the world. Proliferation of bureaucratic controls, extending in the United States to state and local levels, creates a new set of problems in the selection and supervision of controllers. Although some forms of research will be clearly unethical, and already banned under treaties against chemical and biological warfare (CBW), the line often is not clear. Haber and Bosch originally developed their method for military purposes, yet until comparatively recently it was acclaimed as a humanitarian miracle. But what if the order had been reversed? Is it really possible to develop the countermeasures against CBW without gaining knowledge that could be used aggressively? John von Neumann, despite his misjudgements of other issues, argued persuasively that benign and threatening research could not be neatly separated, observing that scientific and technological branches are linked so closely that only the end of technological progress could hold back potentially dangerous knowledge.

Even entirely innocuous applications could lead to surprising social changes. Will nano fabrication be controlled, for example, by a small number of patents? If so, who will hold them and how will they be licensed? Will high fixed costs tend to concentrate production in a few global firms, as in the microchip industry, or will there be opportunity for independents alongside the giants? These questions may seem remote now, but are likely to have consequences as dramatic as the effects of personal computing and the Internet. The benefits of successful nanotechnology could raise living standards globally while creating local crises, for example by slashing the world prices of minerals, gemstones, and other resources that are the mainstays of national and regional economies. Many resource-rich countries are already unstable. And even the United States could be adversely affected; the worldwide spread of nanotechnology processes and skills may reduce the advantages of our agricultural and mineral wealth. In the West, improvements may have paradoxical consequences. If nanotechnologybased medical tests follow the pattern of previous advances, they may increase anxiety and medical costs by producing large numbers of false positives and requiring further testing. (Risk information does not always improve decisions: a recent Swedish study found that parents, informed that their children had inherited genes associated with earlyonset emphysema, were so upset that they began to smoke more rather than quit as the investigators had expected [Brave 2000].) While many cancers and other diseases can be treated effectively if detected early, information about incipient disease or a nontreatable genetic predisposition could actually be a psychological threat to health: "toxic knowledge," some have called it.

Every important new therapy should be reason for rejoicing, but we should not suppose that more effective prevention and treatment will lower medical costs. To the contrary, as health and longevity improve, society pays more for medical care because large numbers of people live to an advanced age and require even more treatment. In 1997, a research group in the Netherlands even found that if all smokers immediately quit, prolonging their lives, medical expenses to society would increase over time. We all hope that nanotechnology will improve the quality as well as the length of life, but because medical advances tend to increase rather than reduce the long-term need for the time of skilled professionals, they are unlikely to be cheap. Already parts of the United States face severe a severe shortage of nurses. Alzheimer's disease threatens to be the epidemic of the twenty-first century, according to some commentators, compounding the need for personnel. Hidden labor costs of technological breakthroughs are not peculiar to medicine. Software for conventional computer processors now requires ever-larger teams of well-paid programmers writing ever-bulkier code. Nanotechnology may accelerate the increase of processor speed and storage capacity, but even with new programming tools, it may require even more programming time. Certainly the explosion of electronic resources in education has tended to increase rather than reduce costs. In all fields, the success of new high-technology methods also has the unintended effect of eroding older skills that remain needed if only as backups. Thus many young physicians now are said to be unable to use a stethoscope properly. Many craft and industrial skills are also declining. Yet in the event of major environmental or social disruption in which high technology no longer functions, these abilities would be crucial resources, as paper-based information would be. In fact some traditional skills remain crucially important even for the most sophisticated technology; experimentation in physics and chemistry still depends on machinists, glassblowers, and other master artisans and technicians.

While many paradoxes accompany radical innovation, opposing it can bring equally strange results. If we freeze technology, we perpetuate and amplify the environmental and social costs of the status quo, including the degradation of air and water quality and the acceleration of climate change. We are on a technological treadmill. We have to find new ways to do things, and nanotechnology can not be excluded. Very possibly, its greatest benefit may be an apparently humble one, like the reduction of rolling friction with a new generation of bearings. There is a precedent: Edwin Mansfield of the University of Pennsylvania determined that innovations in sewing thread had improved the standard of living more than high-technology devices (Weinstein 1993). By permitting faster sewing machine speeds, it had substantially reduced the cost of clothing. And apparently trivial research can have profound results: the chemistry of magnetic tape was an indirect consequence of German research on binders for the gold particles on the Black Russian cigarettes of the 1920s (Fantel 1987). Nanotechnology promises new beginnings. Researchers should bring not only a strong ethical sense but awe at the complexity with which the natural and human worlds interact. They should encourage participation both of lay people and of other professionals familiar with long-term as well as immediate risks: public health specialists, conservation biologists, and environmental historians. No advice can be infallible, but nanotechnology researchers have a rare opportunity to avoid or mitigate the kinds of unintended consequences that have accompanied other major innovations. It would be a great mistake to try to avoid all mistakes. But it would be an even greater error to forget the modesty that the history of technology teaches.

Radically new technologies imply radically new social issues and opportunities. I propose a cultural ecology of nanotechnology in which we find ways to infuse technological development with deeper, more thoughtful and wide-ranging discussions of the social purposes of technology. I chose the ecology metaphor to signify the integration of science and society, to draw attention to interdependencies characteristic of ecologies. (See Nardi and O'Day 1999 on *information ecologies*.) As part of a nanotechnology initiative I would like to see a new science of cost-benefit analysis in which issues of ethics and social responsibility as they relate to technology can be rethought in radical new ways. Perhaps the term "cost-benefit analysis" should be replaced as it connotes limited economic considerations to many. We need to channel energy into the invention of a holistic process of technological development within which we can entertain questions of the human purposes and benefits of technology. The thrust of such an effort would not be prediction or a simplistic notion of managed change, but the development of a new way of approaching our relationship to technology. In a cultural ecology of nanotechnology, we would take seriously the promises of nanotechnology such as cleaner manufacturing, decreased waste, or marvelous medical devices. We would put socially beneficial technologies at the top of the research list. We would find new ways to distribute technologies such as medical devices equitably, we would encourage (somehow) companies to use safer technologies such as, say, nanotech tires that don't fray. Whatever our social desiderata, we would find ways to fuse them to the development and deployment of new technology.

For me, such a process of designing ecologies of technology is desirable because I am not as optimistic as others that technological development always comes out for the best in the long run. Sometimes it does and sometimes it does not. I feel we are currently paying too high a price in pollution, noise, traffic congestion, loss of nature, and lack of safety in our technologies. These poor outcomes are a result of the characteristics of specific technologies coupled with the ways we use the technologies. Traffic gridlock could probably never have happened without the electric self-starter; it is important to remember that specific technologies do matter. On the use side, we needn't have had gridlock had we planned our transportation system differently, insisting on a diverse system of public and private vehicles. If we had gone with the early electric vehicles that were starting to be marketed early in the 20th century, air pollution would be much less of a problem. In a cultural ecology of technology, the relationships between attributes of specific technologies and the ways we actually use the technologies are a key focus of interest. In a cultural ecology of nanotechnology, I'd like to see discussion of how to embed the new technologies into society in radically new, socially beneficial ways. We need to have discussions, for example, about the social contradictions inherent in some of the nanotech promises. How do more long-lasting durable products play with our current economic system based on not-so-durable products that must be replaced often to increase profits for manufacturers? Nanotechnology promises tires that don't fray but we have technologies for safer vehicles now that corporations have sometimes chosen not to use. We know how to manufacture less wastefully, but often we don't do it because it reduces profit. In a cultural ecology of technology such concerns would be a serious topic of discussion and focus of creativity. The government documents promoting nanotechnology that I have read make no mention of the risks of nanotechnology. Are there none? What if we had had cultural ecologies of technology a hundred years ago, and had thought through the implications of having millions of internal combustion engines on the road? Kettering invented the electric selfstarter in 1911. That would have been a great time to undertake a serious envisioning task in which we would have imagined everyone having a car, however fantastic it might have seemed at the time.

A key activity of cultural ecologies of technology is envisionment. Techniques for examining multiple possible scenarios exist now and could be expanded and developed. Could we have envisioned millenial Los Angeles in 1911? Probably not. We didn't know how. We still don't. But we should learn how. Actually, we are already envisioning, but doing a poor job of it. We do not hesitate to conjure wondrous benefits of technology, or, less often, to forecast dystopic visions of technological annihilation. Neither is usually realistic. We need to create new processes to envision both benefits and risks of technology *and the relations between them*. This endeavor in itself is an area for technical creativity. That we have been bad at predicting the future in the past is no reason to avoid this critical task now. If we can talk about creating self-replicating machines out of atoms, we can talk about new techniques for envisioning the consequences of technology. There is no reason we cannot apply our sociological and scientific imaginations to assessing the benefits and risks of technology.

I have noticed that proponents of new technologies often follow a two-part logic in advocating for the development of their technology, however risky it might seem to others. The first part of the argument is the confident prediction of great new benefits. For nanotechnology, we have faster computer chips, very high resolution printers, compact high-volume data storage devices, and new medical technologies such as tiny probes, sensors, drug delivery devices, and ways to regenerate bone and tissue. The second part of the argument, should we pose questions about the potential risks of the technology, is that we don't know how to predict where technology is going. When we stop to ponder potential, even likely, risks of technologies, derisive stories of our poor record of prediction in the past are trotted out. This duality — prediction of fantastic benefits coupled with an assertion that we cannot predict outcomes — is an unhealthy, illogical combination. There's something wrong when the prediction can only be on one side, when we are promised benefits but not allowed to assess risks. Lured by the promises, in which "objective scientific facts" are often invoked as part of the rhetoric of prediction, we go forward, leaving ourselves powerless to envision and prevent negative consequences. The Russian psychologist Vladimir Zinchenko posits something he calls "the ideal form" as a crucial aspect of human social and psychological development (Zinchenko, 1996). The ideal form is where we want to be. Techniques of envisionment are not simply simulations of predicted outcomes, because they contain an element of social purpose. A cultural ecology of technology envisions ideal forms grounded in realistic assumptions, and suggests desirable paths where choices can be made.

A second thing to work out in a cultural ecology is design for co-evolution. To do this, we would give ourselves ample time for discussing and designing how a social and a technical process could co-evolve together gracefully and proactively. Possible venues for such discussions are workshops and programs sponsored by agencies such as the National Science Foundation.

The notion of design for co-evolution resonates with the idea of co-evolution in Brown and Duguid's response to Bill Joy's *Wired* article on the dangers of nanotechnology.

Brown and Duguid point out the strong social influences on technological development. However, I would like to propose that we design and implement a socio-scientific cycle quite different than Brown and Duguid's, which is largely reactive. Our current model of technological development is full speed ahead — and then slam on the brakes when we get scared. Brown and Duguid provide examples of the application of posthoc corrective measures with technologies such as nuclear power. As they point out, it took "the environmental movement, anti-nuclear protests, concerned scientists, worried neighbors of Chernobyl and Three Mile Island, and NIMBY corporate shareholder rebellions to slow the nuclear juggernaut to a manageable crawl." Good grief, do we have call out the scientists, investors, tree huggers, and little old ladies in tennis shoes every time? With nanotechnology, genetically modified foods, cloning, and other technologies with global implications looming, we need a better process. In a cultural ecology of technology, we would be proactive about technological development, not reactive. We would shape technology for our own collectively defined purposes and not confine ourselves to mobilizing to slow dangerous or undesirable juggernauts. And while nuclear power is a hopeful example of co-evolution, it also required the sacrifice of thousands of lives and continuing ill health for many thousands more. While we have successfully held some technologies at bay, other technologies are out of control — such as the automobile. We don't have a healthy ecology for the automobile. Breathing exhaust and spending one's short existence crawling along the freeway is hardly a gift from the gods. It's difficult to say what will happen with genetically modified foods. Despite protests, the brake has not been applied. Half the soybeans in the U.S. and a third of the corn are grown from genetically modified seed. We do not know what ecological or economic effects this will have in the medium to long run. Less threatening but still a kind of daily water torture are technologies such as phone menus that leave people feeling frustrated and diminished. The automated voice response system my bank uses employs poor concatenation technology, stringing together individually recorded numbers that produce a sing-song-y, barely comprehensible voice response to simple requests for account information. These voice systems are especially difficult for people who are hard of hearing or not native speakers of English. Our ecologies often have a *monocultural* character, serving the single need of profit or socalled efficiency. (I sometimes wonder whose efficiency is served as I fight my way through a maze of key presses in a phone menu.) Zinchenko (1996) suggests that two of our most human attributes are creativity and the ability to resist. Brown and Duguid are betting on resistance. This is a time-tested strategy, and one advocated by some of the best minds of our time such as Michel Foucault and Jacques Ellul. But with the rapid pace of technological change, resistance may no longer be sufficient. It's looking to me like Star Trek was right, "Resistance is futile." By the time we mobilize to resist, a lot of damage may have been done. We need to anticipate and plan. My suggestion is to apply human creativity to the problem of designing our technologies in a process that marries the social and the scientific, that treats technology systematically, ecologically. *I believe we can draw on deep wells of creativity that we have not tapped to do this.* Brown and Duguid suggest that Bill Joy's concerns about nanotechnology are distorted by technological tunnel vision. But Joy is far from oblivious to the social. He situates the development of nanotechnology very realistically as a product of "global capitalism and its manifold financial incentives and competitive pressures." Capitalism is a social force more powerful than the "government, courts, formal and informal organizations, social movements, professional networks, local communities, and market institutions" enumerated by Brown and Duguid. Indeed many of these social forms are deeply implicated in capitalism, not outside of it. Forces that can "redirect the raw power of technologies," as Brown and Duguid say, come up against the "manifold financial and competitive pressures" of which Joy speaks. Some parts of government and certain social movements do exist as reactive forces trying to slow and restrain rampant capitalism. However, I believe such forces should be primary generative stimuli of planned societal progress, not catch-up rearguard actions. Joy's fear of "undirected growth" is one to take seriously. Politically experienced people probably find the idea of creating a new social process that intimately links society and science naive and unwieldy. But if we can manufacture devices where a billionth of a meter is a meaningful measure, there is the possibility that we can shape our social processes in just as radical a way.

The importance of communicating advances in nanotechnology to the public will be essential if we are to expect intelligent feedback on our efforts. The question lies in how are we to provide information of such a highly technical nature to a public not yet accustomed to the language of phenomena on the nano level. One way is to use pictures. Images made with a deep and simultaneous respect for the technical, informational and aesthetic content will be important for the communication of nanotechnology to the public. When used with intelligent and accessible text, images of all kinds will play a major role in engaging the interest of the general public. Engagement is the first step in the public's accessibility to information allowing for improved and productive feedback. For general audiences, the photograph is usually less intimidating than the graph and may be one way to permit the viewer to enter into the world of research and to ask questions. There are, of course, a number of other visual representations of data, i.e., animation, illustration, etc. All should be considered part of any investigation of research; especially in nanotechnology. SEEING what are the complex phenomena of nanotechnological research can only encourage expanded interest to the public.

> I believe that we who are privileged to see science's splendor, who image it, diagram it, model it, graph it, and compose its data, can turn the world around, dazzling it with what inspires and nourishes our thinking, if we refine the visual vocabulary we use to communicate our investigations and incorporate — beautifully and above all accurately — the visual component that is already there.

Nanotechnology is a term used to describe a multiplicity of different processes, applications and products which have only one thing in common: characteristic dimensions within the nanometer range. We now have at our disposal materials and technologies such as nanoparticles and nanostructures, manufacturing and analytical techniques, and at the same time an almost unlimited number of applications in which nanotechnologies can be used to improve existing products or create new, previously unattainable product properties. Over the next 20 minutes I will briefly run through the possible uses of nanotechnology and then give a survey of the growth forecasts for the nanotechnology market, while concentrating mainly on the actual materials – particles, composites and coatings.

The cross-sectional significance of nanotechnology clearly emerges from the results of a survey conducted by the American industrial association Nano Business Alliance (NBA). One hundred and fifty nanotechnology start-up companies and spin-offs listed an extensive range of applications impacting almost all areas of life. Nanomaterials and production techniques accounted for a large proportion of almost one third. Nanotechnology is widely regarded as the technology that will be calling the shots in the 21st century. Its dominant position results in substantial and steadily growing government sponsorship volumes in all industrial regions, the lion's share so far coming from the USA. Almost one third of the worldwide state subsidies of about •1.6 billion in 2001 was contributed by the United States. However, this is not reflected in ratings given by technology leaders: when asked about the regional distribution of nanotechnology competence, responders to a survey conducted by 3i, one of the largest, internationally operating venture capital companies, saw Europe as leading the field in pharmaceuticals, materials and chemistry. Concentrated expertise in electronics is ascribed to Japan and Korea. The USA is perceived as the world leader in the nanostructuring field.

Nanotechnology has become the focus of immense expectations in terms of market potential and efficiency. The public perception of this scientific discipline is dominated by images of futuristic nanoproducts such as self-replicating nanoro-bots designed to destroy undesired substances in the blood or tiny, high-performance computers and memories. How big is the gap between fantasy and reality? We can answer this question by taking a closer look at existing market estimates. First let us consider the forecasts for the world market in nanotechnology products: the current volume is estimated at somewhat more than •50 billion (Nano Business Alliance 2001, DG Bank). Assuming mean annual growth rates of 15 to 17 percent, DG Bank anticipates sales of •220 billion for the year 2010. The NBA expects this market volume to be reached in 2005, while in the year 2015 the annual volume of nanotechnology impacted products and services is projected to reach as much as •1,100 billion.

The enormous discrepancy between these two estimates can be traced to several factors:

- It is very important to demarcate between nanotechnology and conventional technologies. DG Bank confines its analysis mainly to materials and systems, while the NBA also includes nanotechnology-influenced products.
- Estimating the net added value of these established products is a problem. Market volume can only be estimated with some degree of reliability for original nanoproducts such as particles.

- Market forecasts are also difficult because nanotechnology has progressed at different rates in different fields. For example, while analytical instruments are already well advanced and have thus been essential to achieving progress in other areas of technology, this is not the case for applications in medicine, pharmacy, biotechnology and machine construction, which are still at an early stage of development.

Let us now examine more closely the structure of current nanotechnology sales of •54 billion. DG Bank divides up this sales volume between nanoparticles and nanocomposites (23 percent), layers (44 percent), systems for analysis of nanostructures (6 percent), surface processing techniques (24 percent) and lateral structures (3 percent), which are unidimensional structures of the kind produced by lithographic processes in chip manufacture. By the year 2010, the world market volume for nanotechnology will have increased more than fourfold. This trend will be driven mainly by new nanotechnology products for which DG Bank anticipates growth rates of 20 to 30 percent, with the main emphasis shifting towards lateral structures, as particles and composites. For the chemical industry, the main focus of interest will be on particles and composites as well as layers. Let's look at the particles first. This segment, which currently accounts for just under one quarter of the world nanotechnology market, is mainly dominated by well-known and proven products of chemical nanotechnology in which BASF has a considerable share with its range of pigment particles, catalysts and consumer products, as well as dispersions for adhesives, coatings and pigments or nanoscale carotinoids. The market share of metal or oxide particles, which determine the public perception of nanoparticles, was estimated by the American research company Business Communication Corporation (BCC) at $493 million in the year 2000. This market volume represents no more than 4 percent of the world market for nanoparticles and composites of around •13 billion, effectively underlining that well-established nanotechnology products presently have ascendancy on the market.

Particles with primary particle sizes below 100 nanometers are characterized by an extremely large surface-to-volume ratio: whereas with particle sizes of 1 micrometer, i.e. 1000 nanometers, only about 1.5 thousandths of a percent of all atoms are located on the surface, with a diameter of 10 nanometers it is already about 15 percent. The properties are thus determined mainly by the behavior of the surface. Possible applications include:

- In powder metallurgy, nanoparticles save energy due to lower sintering temperatures, and higher densities are achieved.
- In cosmetics, titanium dioxide particles are used as UV absorbers in sunscreen creams. These particles do not scatter light in the visible wavelength range, which means that the sun cream is no longer white but transparent. Because of the larger surface-to-volume ratio, however, the UV absorption is better than with conventional particle sizes.

- In lapping substrates for the production of computer chips, memory boards or high performance optical systems, better surface qualities can be achieved with nanoparticles, thereby reducing the reject rate.

Based on a figure of $493 million in the year 2000, the market volume for new products will have almost doubled by 2005 to reach $900 million. This represents an annual growth rate of about 13 percent. While these sales expectations appear realistic, the low level compared to the overall particles and composites segment should be remembered. Most of today's commercial nanoparticles consist of silicon dioxide, titanium dioxide, aluminum oxide and iron oxide. BCC expects the volume of nanoscale silicon dioxide to increase almost threefold by 2005 to reach a market share of almost 44 percent. This is plausible in view of the broad range of applications for the comparatively low-cost silicon dioxide.

Now let us survey the main areas of application for inorganic nanoparticles: According to BCC, the main purchasers of these particles are the electronics and IT industries which by the year 2005 will require almost three quarters of the total production. The biggest single application in this segment is CMP slurries (CMP, chemical mechanical planarization). These are powder suspensions for chemical-mechanical surface lapping and polishing of silicon wafers. Mainly silicon dioxide and aluminum oxide slurries are used in this process. The second largest application for nanoparticles is magnetic storage media. In 2005, the share of biomedicine and pharmaceutics is expected to reach 16 percent, with about 10 percent going for energy, catalysis and structural applications.

Small firms are a major source of innovations in new materials. High growth rates are therefore expected for start-up companies which currently have little or no sales. Many applications serve niches, but some have the potential to generate high sales volume. As a comparison: the market for titanium dioxide particles as inorganic UV absorbers in cosmetics, as already mentioned, is currently estimated at about $140 million while in the year 2000 the world market for biomarkers reportedly reached $4 billion (Thetareports, BJP Publications Inc., UK, 2000). Holding companies of BASF Venture Capital GmbH are already active in these future-oriented segments. The direct investment company Oxonica, Oxford, UK, for example, is developing UV absorbers for cosmetics, fluorescent nanoparticles as biomarkers and nanoparticles for catalysis. A holding company of the NextGen Enabling Technologies Fund headquartered in California, in which BASF Venture Capital GmbH is an active limited partner, is the company Catalytic Solutions Inc., Oxnard, California. This company succeeded in significantly reducing the content of noble metals in automobile catalytic converters. These products are also much cheaper and longerlasting than conventional systems. Customers include the Honda Motor Company and the automobile components supplier Car Sound Exhaust Systems. The efficiency of the products developed by Catalytic Solutions is so good that established manufacturers of automobile catalytic converters were downrated by the analysts due to the superior CSI technology.

Another potential use for nanoparticles is as fillers in plastics. These materials, known as nanocomposites, exhibit several advantages compared to conventionally filled plastics. For example, improvements have been achieved in mechanical behavior, barrier properties or processing characteristics. This gives rise to a number of applications, such as bottles for carbonated beverages, antistatic packagings for sensitive electronic equipment or automotive components with reduced wall thickness and thus lower weight. But there are also disadvantages, such as,

- compounding problems due to the homogeneous filler distribution,
- the often observed tendency of composites to become brittle and, especially,
- the high price of many fillers. The cost/performance ratio is not competitive.

Turnover with nanocomposites in the year 2001 is estimated as equivalent to about •300 million (Plastic News, assuming an average price of •3 per kilogram). Various sources (Chemical Business Newsbase, Plastic News), however, anticipate a marked upturn in sales to the equivalent of as much as •1.5 billion by the year 2009, representing 500,000 metric tons per year. We consider these annual growth rates of about 100 percent to be overly optimistic, since improvements in properties have so far lagged behind the high expectations and the prices of fillers are still too high for large-scale applications. According to Bins & Associates, an American market research company located in Sheboygen, Wisconsin, the market will be dominated by nanoclay composites containing nanoscale clay plates as filler. Other nanofillers such as calcium carbonate or carbon nanotubes, on the other hand, will take a back seat. This is consistent with our own expectation that nanoclay will set the pace as a very competitively priced filler. European Chemical News expects to see high growth rates by the year 2009, especially in the construction sector but also in the packaging and automotive industries. BASF is also active in these fields. Basell, a joint venture of BASF and Shell, for example, supplies a polypropylene reinforced with nanoscale clay for the footrest of a GM minivan. The improved mechanics allow the weight to be reduced by just under 20 percent. Now we come to another area of nanotechnology of interest to us, namely coatings, which with 44 percent account for a large share of the overall nanotechnology market of •54 billion.

German industry leads the field in ultrathin coatings and the associated coating technology. Major contributors to this success include the Institute for New Materials in Saarbrücken and the Fraunhofer Institutes in Freiburg and in Würzburg. The world market for coatings was estimated at about •24 billion in 2001. DG Bank anticipates an increase to •81 billion by the year 2010, representing an annual growth rate of 14 percent. According to the German Association of Engineers (VDI), the established applications in electronics and information technology provided the lion's share of sales in 2001 with about 56 percent: this includes coatings for data storage media (•8 billion) and thin layers for information technology components (about •6 billion).

The VDI estimates the volume for hard coatings in 2001 at around •3.5 billion. Also traditionally strong is the optical coating of glass and plastics at around •900 million. As with nanoparticles, a substantial proportion of sales of coatings is also generated witih established applications.

Above average growth rates are anticipated for innovative functional coatings. Examples of such novel coating systems are easy-to-clean or even self-cleaning surfaces. These include the lotus effect already mentioned in a previous presentation, which is being progressed as a project by BASF Future Business. Even further removed from actual implementation are smart surfaces with switchable properties which, for example, can change their color or adhesiveness or photoelectric surface coatings which allow systems (such as air conditioning) to be operated without additional energy input.

Another example is innovative optical coatings for non-reflecting surfaces and scratch protection. A window on technology into this field has been opened for us by the participation of NextGen Enabling Technologies in the nanotechnology start-up company Optiva, San Francisco. Optiva is developing products such as thin-layer polarizers without which LCD flat displays do not function. They are much cheaper than conventional technologies and also feature a number of improved properties such as lower thickness, greater strength and high efficiency. Let me summarize as follows:

- Nanotechnology as a cross-sectional technology will play an important future role in almost all areas of technology.
- However: in the year 2001, more than 85 percent of sales with nanomaterials and coatings was still generated by the sale of established products.
- Forecasting market trends is difficult and in our view liable to excessive euphoria, firstly because some nanotechnologies are still at a very early stage of development, and secondly because forecasts are based on difficult-to-estimate sales shares for products manufactured by nanotechnology.
- The biggest obstacles to conquering the nanomarket are:
  - Nanoresearch is frequently technology driven. Too much attention is paid to the fascinating dimensions and too little to the property profile.
  - Nanotechnology based solutions are often too expensive and come off second best when compared to the cost/performance ratio of other solutions. This leaves us with the following conclusions:
- Earning money with nanotechnology is much more difficult than many people believe in their euphoria.
- Even in ten years' time, a very large proportion of nanotechnology based sales will still be generated by classical nanotechnology, in other words with pigments and dispersions.

BASF is well positioned in the operational field "nanotechnology":

- We already have an established business in applications such as dispersions and color pigments and are participating in their continued growth.
- We have also created innovative instruments in the form of new business development units, BASF Future Business GmbH and BASF Venture Capital GmbH, which give access to new developments and participation in the business conducted in this sector.

REFERENCES

Attebery, B.: 2004, 'Dust, Lust, and Other Messages from the Quantum Wonderland', in: N.K. Hayles (ed.), *Nanoculture: Implications of the New Technoscience*, Bristol, UK: Intellect Books, pp. 161-172.

Bensaude-Vincent, B.: 2001, 'The Construction of a Discipline: Materials Science in the United States', *Historical Studies in the Physical and Biological Sciences*, 31, 223-248.

Bensaude-Vincent, B.: 2004, 'Two Cultures of Nanotechnology?', *Hyle: International Journal for Philosophy of Chemistry*, 10(2), 65-82.

Bueno, O.: 2004, 'The Drexler-Smalley Debated on Nanotechnology: Incommensurability at Work?', *Hyle: International Journal for Philosophy of Chemistry*, 10(2), 83-98.

Bueno, O.: 2004, 'Von Neumann, Self-Reproduction and the Constitution of Nanophenomena', in: D. Baird, A. Nordmann & J. Schummer (eds.), *Discovering the Nanoscale*, Amsterdam: IOS Press, pp. 101-115.

Carroll, J.S.: 2001, 'Social Science Research Methods for Assessing Societal Implications of Nanotechnology', in: M.C. Roco and W.S. Bainbridge (eds.), *Societal Implications of Nanoscience and Nanotechnology*, Dordrecht: Kluwer, pp. 188-192.

Chen, C-Y., S.L. Burkett, H.-X. Li, and M.E. Davis. 1993. *Microporous Mater*. 2:27.

Chianelli, R.R. 1998. Synthesis, fundamental properties and applications of nanocrystals, sheets, and fullerenes based on layered transition metal chalcogenides. In *R&D status and trends*, ed. Siegel et al.

Chianelli, R.R., M. Daage, and M.J. Ledoux. 1994. *Advances in Catalysis* 40:177.

Claeson, T., and Likharev, K.," Single Electronics," *Scientific American*, June 1992, pp. 80-85.

Coenen, C.: 2004, 'Nanofuturismus: Anmerkungen zu seiner Relevanz, Analyse und Bewertung', *Technikfolgenabschätzung ' Theorie und Praxis*, 13 (2), 78-85.

Crandall, B.C. and Lewis, J., *Nanotechnology: Research and Perspectives*, MIT Press, Cambridge, 1992.

Crow, M. & Sarewitz, D.: 2001, 'Nanotechnology and Societal Transformation', in: M.C. Roco and W.S. Bainbridge (eds.), *Societal Implications of Nanoscience and Nanotechnology*, Dordrecht: Kluwer, pp. 45-54.

de Souza e Silva, A.: 2004, 'The Invisible Imaginary: Museum Spaces, Hybrid Reality and Nanotechnology', in: N.K. Hayles (ed.), *Nanoculture: Implications of the New Technoscience*, Bristol, UK: Intellect Books, pp. 27-82.

Decker, M.; Fiedeler, U.; Fleischer, Th.: 2004, 'Ich sehe was, was Du nicht siehst... zur Definition von Nanotechnologie', *Technikfolgenabschätzung ' Theorie und Praxis*, 13 (2), 10-16.

Grinbaum, A. & Dupuy, J.-P.: 2004, 'Living with Uncertainty: Toward the Ongoing Normative Assessment of Nanotechnology', *Techne: Research in Philosophy and Technology*, 8(3) (forthcoming).

Grinbaum, A.: 2004, 'La condition de l'homme moderne et les nanotechnologies', in: G. Nivat (ed.), *Les limites de l'humain. 39èmes Rencontres Internationales de Genève*, L'Age d'Homme, Genève, p. 141.

Johansson, M.: 2003, 'Plenty of room at the bottom: Towards an anthropology of nanoscience', *Anthropology Today*, 19 (No. 6), 3-6.

\Kresge, C.T., M.E. Leonowicz, W.J. Roth, J.C. Vartuli, and J.S. Beck. 1992. *Nature* 359:710.

Kupperman, A., S. Nadimi, S. Oliver, G. Ožin, J. Garcés, and M. Olken. 1993. *Nature* 365:239.

Kuusi, O.; Meyer, M.: 2002, 'Technological generalizations and leitbilder ' the anticipation of technological opportunities', *Technological Forecasting & Social Change*, 69, 625-639.

Landon, B.: 2004, 'Less is More: Much Less is Much More: The Insistent Allure of Nanotechnology Narratives in Science Fiction', in: N.K. Hayles (ed.), *Nanoculture: Implications of the New Technoscience*, Bristol, UK: Intellect Books, pp. 131-146.

Laszlo, P.: 2004, 'Is There Life After Partington?', *Hyle: International Journal for Philosophy of Chemistry*, 10(2), 169-178.

Lenhard, J.: 2004, 'Nanoscience and the Janus-Faced Character of Simulations', in: D. Baird, A. Nordmann & J. Schummer (eds.), *Discovering the Nanoscale*, Amsterdam: IOS Press, pp. 93-100.

Lent, C.S., Tougaw, P.D., Porod, W., Bernstein, G.H., "Quantum Cellular Automata," *Nanotechnology*, Vol. 4, p. 49, 1993.

Lewak, S.E.: 2004, 'What's the Buzz? Tell Me What's A-Happening: Wonder, Nanotechnology, and Alice's Adventures in Wonderland', in: N.K. Hayles (ed.), *Nanoculture: Implications of the New Technoscience*, Bristol, UK: Intellect Books, pp. 201-.

Lide, D.R., ed. 1993-1994. *CRC Handbook of Chemistry and Physics*, 74th ed.

Lin-Easton, P.C.: 2001, 'It's Time for Environmentalists to Think Small ' Real Small: A Call for the Involvement of Environmental Lawyers in Developing Precautionary Policies Molecular Nanotechnology', *Georgetown International Law Review*, 14, 106-134.

López, J.: 2004, 'Bridging the Gaps: Science Fiction in Nanotechnology', *Hyle: International Journal for Philosophy of Chemistry*, 10(2), 129-152.

Lösch, A.: 2004, 'Nanomedicine and Space: Discursive Orders of Mediating Innovations', in: D. Baird, A. Nordmann & J. Schummer (eds.), *Discovering the Nanoscale*, Amsterdam: IOS Press, pp. 193-202.

Marshall, K.: 2004, 'Atomizing the Risk Technology', in: N.K. Hayles (ed.), *Nanoculture: Implications of the New Technoscience*, Bristol, UK: Intellect Books, pp. 147-160.

Martin, T.P., N. Malinowski, U. Zimmerman, U. Naher, and H. Schaber. 1993. *J. Chem. Phys.* 99:4210.

Martin, T.P., U. Naher, H. Schaber, U. Zimmerman. 1993. *Phys. Rev. Lett.* 70:3079.

Mayer, S.: 2002, 'From genetic modification to nanotechnology: the dangers of 'sound science'', in:

T. Gilland (ed.), *Science: Can We Trust the Experts?*, London: Hodder and Stoughton, pp. 1-15.

Montemerlo, M.S., Love, J.C., Opiteck, G.J., Goldhaber, D. J., and Ellenbogen, J.C., "Technologies and Designs for Electronic Nanocomputers," MITRE Technical Report 96W0000044, The MITRE Corporation, McLean, VA, July 1996. For more information, send e-mail to nanotech@mitre.org.

Moor, J.H. & Weckert, J.: 2004, 'Nanoethics: Assessing the Nanoscale From an Ethical Point of View', in: D. Baird, A. Nordmann & J. Schummer (eds.), *Discovering the Nanoscale*, Amsterdam: IOS Press, pp. 301-310.

Prigogine, I., and S. Rice. 1988. *Advances in chemical physics*, Vol. 70, Parts 1 & 2. New York: J. Wiley.

R. T. Bate, "Nanoelectronics," *Nanotechnology*, Vol. 1, pp. 1-7, 1990.

Rademann, K., B. Kaiser, U. Even, F. Hensel. 1987. *Phys. Rev. Lett.* 59:2319.

Rao, C.N.R., B.C. Satishkumar, and A. Govindaraj. 1997. *Chem. Commun.* 1581.

Rao, M.B., and S. Sircar. 1993. *Gas Separation and Purification* 7:279.

Reed, M.A., "Quantum Dots," *Scientific American,* January 1993, pp. 118-123.

Roberts, J.A.: 2004, 'Deciding the Future of Nanotechnologies: Legal Perspectives on Issues of Democracy and Technology', in: D. Baird, A. Nordmann & J. Schummer (eds.), *Discovering the Nanoscale*, Amsterdam: IOS Press, pp. 247-255.

Robinson, C.: 2004, 'Images in NanoScience/Technology', in: D. Baird, A. Nordmann & J. Schummer (eds.), *Discovering the Nanoscale*, Amsterdam: IOS Press, pp. 165-169.

Robison, W.L.: 2004, 'Nano-Ethics', in: D. Baird, A. Nordmann & J. Schummer (eds.), *Discovering the Nanoscale*, Amsterdam: IOS Press, pp. 285-300.

Roco, M.C. & Bainbridge, W.S. (eds.): 2001, *Societal implications of nanoscience and nanotechnology*, (Proceedings of a workshop organized by the National Science Foundation, September 28-29, 2000), Kluwer: Dordrecht [available online at http://itri.loyola.edu/nano/societalimpact/nanosi.pdf]

Roco, M.C. & Tomellini, R. (eds.): 2002, *Nanotechnology: Revolutionary Opportunities and Societal Implications, Workshop, Lecce (Italy), 31 January - 1 February 2002*, Luxemburg: Office for Official Publications of the European Communities, [200 pp.]

Roco, M.C.; Bainbridge, W.S. (eds.): 2002, *Converging Technologies for Improving Human Performance: Nanotechnology, Biotechnology, Information Technology and the Cognitive Science*, Arlington, VA: National Science Foundation.

Roher, H. 1993. *Jpn. J. Appl. Phys.* 32:1335.

Rohlfing, E.A., D.M. Cox, and A. Kaldor. 1984. *J. Chem. Phys.* 81:3846.

Rosen, A. 1998. A periodic table in three dimensions: A sightseeing tour in the nanometer world. In *Advances in quantum chemistry*. In press.

Ruthven, D.M., S. Farcoq, K.S. Knaebel. 1994. *Pressure swing adsorption*. New York: VCH Publishers.

Sarewitz, D.; Woodhouse, E.: 2003, 'Small is Powerful', in: A. Lightman, D. Sarewitz & Chr. Desser, (eds.), *Living with the Genie: Essays on Technology and the Quest for Human Mastery*, Washington, DC: Island Press, pp. 63-83.

Schiemann, G.: 2004, 'Dissolution of the Nature-Technology Dichotomy? Perspectives on Nanotechnology from an Everyday Understanding of Nature', in: D. Baird, A. Nordmann & J. Schummer (eds.), *Discovering the Nanoscale*, Amsterdam: IOS Press, pp. 209-213.

Schmidt, J.C.: 2004, 'Unbounded Technologies: Working Through Technological Reductionism of Nanotechnology', in: D. Baird, A. Nordmann & J. Schummer (eds.), *Discovering the Nanoscale*, Amsterdam: IOS Press, pp. 35-50.

Siegel, R.W., E. Hu, and M.C. Roco, eds. 1998. *R&D status and trends in nanoparticles, nanostructured materials, and nanodevices in the United States*. Baltimore: Loyola College, International Technology Research Institute. NTIS #PB98-117914.

Stix Gary., "Toward 'Point One'," *Scientific American*, February 1995, pp. 90-95.

Stuckless, J.T, D.E. Starr, D.J. Bald, C.T. Campbell. 1997. *J. Chem. Phys.* 107:5547.

Suchman, M.: 2001, 'Envisioning Life on the Nano-Frontier', in: M.C. Roco and W.S. Bainbridge (eds.), *Societal Implications of Nanoscience and Nanotechnology*, Dordrecht: Kluwer, pp. 211-216.

Suchman, M.: 2002, 'Social Science and Nanotechnologies', M. Roco & R. Tomellini (eds.), *Nanotechnology ' Revolutionary Opportunities and Societal Implications*, Luxembourg: European Communities, pp. 95-99.

Sweeney, A.E.; Seal, S. & Vaidyanathan, P.: 2003, 'The promises and perils of nanoscience and nanotechnology: Exploring emerging social and ethical issues', *Bulletin of Science, Technology & Society*, 23 (4), 236-245.

Tschope, A.S., W. Liu, M. Flyzani-Stephanopoulos, and J.Y. Ying 1995. *J. Catal.* 157:42.

Turton, R., *The Quantum Dot*, Oxford University Press, 1995.

Verbeek, A. & Callaert, J.: 2002, *Detailed analysis of the science-technology interaction in the field of Nanotechnology*, Luxemburg: Office for Official Publications of the European Communities (Linking Science to Technology. Bibliographic References in Patents, Vol. 9).

Vermaas, P.E.: 2004, 'Nanoscale Technology: A Two-Sided Challenge for Interpretations of Quantum Mechanics', in: D. Baird, A. Nordmann & J. Schummer (eds.), *Discovering the Nanoscale*, Amsterdam: IOS Press, pp. 77-91.

Glossary

Active shield: A defensive system with built-in constraints to limit or prevent its offensive use.

Active site: Region of an enzyme surface to which a substrate molecule binds in order to undergo a catalyzed reaction.

Avogadro's number: The number of atoms in exactly 12 grams of pure ^{12}C, equal to 6.022×10^{23}.

Biocompatible coated materials: Biocompatible materials usually used in dental and bone implants that enhance biologic fixation, thereby increasing the bond strength between the coated material and bone, and minimize possible biological effects that may result from the implant itself. MeSH, 1999

Breakdown: Failure of a material resulting from an electrical overload. The resulting damage may be in the form of thermal damage (melting or burning) or electrical damage (loss of polarization in piezoelectric materials). The failing of a physical principle or unit of measurement as the scale is changed.

Brownian Assembly: Brownian motion in a fluid brings molecules together in various position and orientations. If molecules have suitable complementary surfaces, they can bind, assembling to form a specific structure. Brownian assembly is a less paradoxical name for self-assembly (how can a structure assemble itself, or do anything, when it does not yet exist?).

Brownian Motion: Motion of a particle in a fluid owing to thermal agitation, observed in 1827 by Robert Brown. (Originally thought to be caused by vital force, Brownian motion in fact plays a vital role in the assembly and activity of the molecular structures of life).

Bus: Transmission medium for electrical or optical signals that perform a particular function, such as computer control.

Calibration: A process of adapting a sensor output to a known physical quantity to improve sensor output accuracy. Ubiquitous Ca^{2+}-binding protein whose binding to other proteins is governed by changes in intracellular Ca^{2+} concentration. Its binding modifies the activity of many target enzymes and membrane transport proteins.

Carbonyl: A chemical moiety consisting of O with a double bond to C. If the C is bonded to N, the resulting structure is termed an amide; if it is bonded to O, it is termed a carboxylic acid or an ester linkage.

Carboxylic acid: A molecule that includes a C having a double bond to O and a single bond to OH.

Compliance: The reciprocal of stiffness; in a linear elastic system, displacement equals force times compliance.

Copolymer: A polymer that consists of two or more dissimilar monomer units in combination in its molecular chains. Also a polymer formed from the polymerization of more than one type of monomer.

Corrosion: Deteriorative loss of a metal as a result of dissolution environmental reactions.

Cross-linking: A process forming chemical bonds between two separate molecular chains.

Cross-sensitivity: The influence of one measurand on the sensitivity of a sensor, another measurand.

Crosstalk: Electromagnetic noise transmitted between leads or circuits in close proximity to each other.

Crybiology: The science of biology at low temperatures; research in cryobiology has made possible the freezing and storing of sperm and blood for later use.

Crystal Lattice: The regular three-dimensional pattern of atoms in a crystal.

Crystal structure: For crystalline materials, the manner in which atoms or ions are arrayed in space. It is defined in terms of the unit cell geometry and the atom positions within the unit cell.

Crystallescence: In medical nanorobotics, the crystallization of solid solute that is offloaded by nanorobot sorting rotors at a concentration that exceeds the solvation capacity of the surrounding solvent.

Curie temperature (also Curie point) (T_c): The temperature above which a ferromagnetic or ferrimagnetic material becomes paramagnetic. For iron the Curie point is 760oC and for nickel 356°C.

Current [A]: Measure of rate of flow of electric charge: a one-ampere current is a flow of 1 C of charge per second.

Cutoff: Condition in a diode or bipolar junction transistor in which the potential across a p-n junction prevents current flow.

Cyclic: A structure is termed cyclic if its covalent bonds form one or more rings.

Cycloaddition: A reaction in which two unsaturated molecules (or moieties within a molecule) join, forming a ring.

Cyclotron: A type of particle accelerator in which an ion introduced at the center is accelerated in an expanding spiral path by use of alternating electrical fields in the presence of a magnetic field.

Cytocarriage: In medical nanorobotics, the commandeering of a natural motile cell, by a medical nanorobot, for the purposes of in vivo transport (of the nanorobot), or to perform a herding function (of the affected cell), or for other purposes.

Cytocide: The killing of living cells.

Cytography: A physical description (and mapping) of the living cell.

Cytoidentification: Identification of cell type.

Cytometrics: The quantitative measurement of cell sizes, shapes, structures, and numbers.

Cytopenetration: In medical nanorobotics, entry into cells by penetrating the plasma membrane.

Deoxyribonucleic acid (DNA): A huge nucleotide polymer having a double helical structure with complementary bases on two strands. Its major functions are protein synthesis and the storage and transport of genetic information.

Deoxyribonucleic acid: Deoxyribonucleic acid are long chains consisting of four kinds of nucleotides; the order of these nucleotides encodes the information needed to construct protein molecules. These in turn make up much of the molecular machinery of the cell. DNA is the genetic material of cells.

Electrical breakdown: Condition in which, particularly with high electric field, a nominal insulator becomes electrically conducting.

Electromotive force (emf) series: A series of chemical elements arranged in order of their electromotive force. The electromotive force is the greatest potential difference that can be generated by a particular source of electric current. In practice this potential may be observable only when the source is not supplying current, because of its internal resistance.

Ferromagnetism: Permanent and large magnetizations found in some metals (e.g. Fe, Ni and Co), resulting from the parallel alignment of neighboring magnetic moments.

Fertilization: Fusion of a male and a female gamete (both haploid) to form a diploid zygote, which develops into a new individual.

FET: Field-Effect Transistor — semiconductor device whose insulated gate electrode controls current flow.

Fiber-optic: Relates to transmission of information as modulated light in tiny transparent fibers instead of copper wires.

Fibrin sealant: A biological tissue glue derived from human plasma. Sometimes referred to as fibrin glue, it contains the components necessary for the last step of coagulation: i) fibrinogen and other coagulation factors and ii) thrombin. It is used in surgical procedures to arrest bleeding and as an adjunct to wound healing.

Fibrin: A protein derived from FIBRINOGEN in the presence of THROMBIN, which forms part of the blood clot. [MeSH]

Foglet: A mesoscale machine. A discreet component of utility fog. [J. S. Hall 1994]

Follicle cell: One of the cell types that surround a developing oocyte or egg.

Follower: A component in a cam system that is driven through a pattern of displacements as it rests against a moving contoured surface.

Free Radicial: A molecule containing an unpaired electron, typically highly unstable and reactive. Free radicals can damage the molecular machinery of biological systems, leading to cross-linking and mutation.

Free-energy change (DG): Change in the free energy during a reaction: the free energy of the product molecules minus the free energy of the starting molecules. A large negative value of DG indicates that the reaction has a strong tendency to occur.

Frequency response: Two relations between sets of inputs and outputs. One relates frequencies to the output-input amplitude ratio; the other relates frequencies to the phase difference between the output and input.

Frequency: Number of times per second that a quantity representing a signal, such as a voltage, changes state. Also, the number of waves (cycles) per second that pass a given point in space.

G1 phase: Gap 1 phase of the eucaryotic cell-division cycle, between the end of cytokinesis and the start of DNA synthesis.

G2 phase: Gap 2 phase of the eucaryotic cell-division cycle, between the end of DNA synthesis and the beginning of mitosis.

GAG (glycosaminoglycan): Long, linear, highly charged polysaccharide composed of a repeating pair of sugars, one of which is always an amino sugar. Mainly found covalently linked to a protein core in extracellular matrix proteoglycans. Examples include chondroitin sulfate, hyaluronic acid, and heparin.

Gain: The ratio of the amplitude of an output to input signal.

Galvanic corrosion: The preferential corrosion of the more chemically active of two metals electrically coupled and exposed to an electrolyte.

Gamete: Specialized haploid cell, either a sperm or an egg, serving for sexual reproduction.

Ganglion (plural ganglia): Cluster of nerve cells and associated glial cells located outside the central nervous system.

Ganglioside: Any glycolipid having one or more sialic acid residues in its structure. Found in the plasma membrane of eucaryotic cells and especially abundant in nerve cells.

Gap junction: Communicating cell-cell junction that allows ions and small molecules to pass from the cytoplasm of one cell to the cytoplasm of the next.

Gastrula: Animal embryo at an early stage of development where cells are invaginating to form the rudiment of a gut cavity. (From Greek *gaster*, belly.)

Gate: Circuit whose logical output variables are determined by its inputs. In digital logic, a component that can switch the state of an output dependent on the states of one or more inputs.

Heat: As defined in thermodynamics, heat is the energy that flows between two systems as a result of temperature differences (a system *contains* neither heat nor work, but can *produce* heat or *do* work). Heat thus differs from thermal energy.

Heisenberg Uncertainty Principle: A quantum-mechanical principle with the consequence that the position and momentum of an object cannot be precisely determined. The Heisenberg principle helps determine the size of electron clouds, and hence the size of atoms.

HeLa cell: Line of human epithelial cells that grows vigorously in culture. Derived from a human cervical carcinoma.

Heme: An iron complex. Cyclic organic molecule containing an iron atom that carries oxygen in hemoglobin and carries an electron in cytochromes.

Hemidesmosome: Specialized cell junction between an epithelial cell and the underlying basal lamina.

Hemoglobin: A biomolecule composed of four myoglobine-like units (proteins plus heme) that can bind and transport four oxygen molecules in the blood.

The major protein in red blood cells that associates with O_2 in the lungs by means of a bound heme group.

HEMT: High electron mobility transistor.

Henry (H): Unit of inductance (see"inductance"). One henry (H) is the inductance of a closed circuit in which an electromotive force of 1 volt is produced when the electric current in the circuit varies uniformly at the rate of 1 ampere per second.

Henry's law: The amount of gas dissolved in a solution is directly proportional to the pressure of the gas above the solution.

Heterocaryon: Cell with two or more nuclei produced by the fusion of two or more different cells.

Heterochromatin: Region of a chromosome that remains unusually condensed and transcriptionally inactive during interphase.

Highest Occupied Molecular Orbital (HOMO): The highest energy molecular orbital of an atom or molecule that contains an electron.

Histone: A protein present in the eukaryotic nucleus that is bound to the DNA at regularly spaced intervals.

Homogeneous and Heterogeneous Assays: A homogeneous assay does not require a separation step to remove free antigen from bound antigen and relies upon the fact that the function of the label is modified upon binding, leading to a change in signal intensity. Because of high background signal a heterogeneous approach incorporating a separation step of bound and unbound makes the detection limit lower, approaching the values obtained by RIA. The homogeneous assay is less technically demanding.

Human Genome Project: A research initiative that has mapped the entire human genome.

Hysteresis: The difference in the output when a specific input value is approached first with an increasing and then with a decreasing input. This phenomenon occurs in ferroelectric materials and results in irreversible loss of energy through heat dissipation.

Imaging Contrast Agent: A molecule or molecular complex that increases the intensity of the signal detected by an imaging technique, including MRI and ultrasound. An MRI contrast agent, for example, might contain gadolinium attached to a targeting antibody. The antibody would bind to a specific target—a metastatic melanoma cell, for example – while the gadolinium would increase the magnetic signal detected by the MRI scanner.

Imaging Contrast Agent: A molecule or molecular complex that increases the intensity of the signal detected by an imaging technique, including MRI and ultrasound. An MRI contrast agent, for example, might contain gadolinium attached to a targeting antibody. The antibody would bind to a specific target—a metastatic melanoma cell, for example – while the gadolinium would increase the magnetic signal detected by the MRI scanner.

Immortalization: Production of a cell line capable of an unlimited number of cell divisions. Can be the result of a chemical or viral transformation or of fusion with cells of a tumor line.

Immune Machines: Medical nanomachines designed for internal use, especially in the bloodstream and digestive tract, able to identify and disable intruders such as bacteria and viruses.

Immune system: Population of lymphocytes and other white blood cells in the vertebrate body that defends it against infection.

Immunoglobulin (Ig): An antibody molecule. Higher vertebrates have five classes of immunoglobulin - IgA, IgD, IgE, IgG, and IgM - each with a different role in the immune response.

Immunoglobulin like (Ig-like) domain: Characteristic protein domain of about 100 amino acids that is found in antibody molecules and in many other proteins that form the Ig superfamily.

IMP: Electronic implant, especially in the brain. [Ron Hale Evans]

Impedance: The complex ratio of a force-like quantity (force, pressure, voltage, temperature, or electric field) to a corresponding related velocity-like quantity (velocity, volume velocity, current, heat flow, or magnetic field strength).

Inductance [in Henry, H]: That property of an electric circuit which tends to oppose change in current in the circuit. One henry (H) is the inductance of a closed circuit in which an electromotive force of 1 volt is produced when the electric current in the circuit varies uniformly at the rate of 1 ampere per second.

Induction (embryonic): Change in the developmental fate of one tissue caused by an interaction with another tissue.

Inductor: Energy storage circuit component consisting of a coil of wire and possibly a magnetic material.

Inflammatory response: Local response of a tissue to injury or infection. Caused by invasion of white blood cells, which release various local mediators such as histamine.

Infrared: Invisible electromagnetic radiation having a longer wavelength, and lower frequency, than visible red light.

Inhibitor: A chemical substance that, when added in relatively low concentrations, slows down a chemical reaction.

Inline Universities: (As opposed to online universities), nanocomputer implants serving to increase intelligence and education of their owners, essentially turning them into walking universities.

Inmessaging: In medical nanorobotics, conveyance of information from a source external to the human body, or external to working nanodevices, to a receiver located inside the human body.

Insertion point (in lithography context): Adaptation of a new lithography technique is referred to as the insertion point of that technique.

Insulator: A substance, object or material that does not conduct electricity. Material that conducts electricity very poorly.

Integrated Circuit (IC): An electronic circuit consisting of many interconnected devices on one piece of semiconductor, typically into 10 millimeters on a side. ICs are the major building blocks of today's computers. Semiconductor circuit, typically on a very small silicon chip, containing microfabricated transistors, diodes, resistors, capacitors, etc.

Interagency Working Group on Nanoscience, Engineering and Technology IWGN, National Science and Technology Council NSTC, US

Interface: In physical terms, an interface is the boundry between two phases, for instance between a solid and liquid or between a liquid and gas.

Intermolecular: Between more than one molecule. Describes an interaction (e.g., a chemical reaction) between different molecules.

Internal energy: The sum of the kinetic and potential energies (including electromagnetic field energies) of the particles that make up a system.

Internet: Worldwide digital communication network in which packets of information travel between senders and recipients.

Interstitial diffusion: a diffusion mechanism that causes atomic motion from interstitial site to interstitial site.

Ketone: Organic molecule containing a carbonyl group linked to two alkyl groups.

Kevlar (TM): A synthetic fiber made by E. I. du Pont de Nemours & Co., Inc. Stronger than most steels, Kevlar is among the strongest commercially available materials and is used in aerospace construction, bulletproof vests, and other applications requiring a high strength-to-weight ratio.

Kilobyte (kB): 2^{10} (= 1024, or about one thousand) bytes of information.

Kilocalorie (kcal): Unit of heat energy equal to 1000 calories. Often used to express the energy content of food or molecules: bond strengths, for example, are measured in kcal/mole. An alternative unit in wide use is the kilojoule, equal to 0.24 kcal.

Kilohertz (kHz): One thousand cycles per second.

Kilojoule: Standard unit of energy equal to 1000 joules, or 0.24 kilocalories.

Kinesin: One type of motor protein that uses the energy of ATP hydrolysis to move along a microtubule.

Kinetic energy: Energy resulting from the motion of masses.

Kinetic molecular theory: A model that assumes that an ideal gas is composed of tiny particles (molecules) in constant motion.

Kinetochore: Complex structure formed from proteins on a mitotic chromosome to which microtubules attach and which plays an active part in the movement of chromosomes to the pole. The kinetochore forms on the part of the chromosome known as the centromere.

Knowbots: Knowledge robots, first developed Vinton G. Cref and Robert E. Kahn for National Research Initiatives. Knowbots are programmed by users to scan networks for various kinds of related information, regardless of the language or form in which it expressed. "Knowbots support parallel computations at different sites. They communicate with one another, and with various servers in the network and with users."

Mask: Pattern on glass, like a photographic negative, for producing integrated-circuit elements on semiconductor wafer.

Megahertz (Mhz): One million cycles per second

NMOS: An acronym for n-*channel metal-oxide-semiconductor*, as in NMOS transistor and NMOS logic.

Pitting: A form of very localized corrosion wherein small pits or holes form, usually in a vertical direction.

pK value: A measure of the strength of an acid on a logarithmic scale. The pK value is given by log10 (1/Ka), where Ka is the acid dissociation constant pK values often are used to compare the strengths of different acids.

Plastic deformation: Permanent or nonrecoverable deformation, accompanied by permanent atomic displacement.

Resonant frequency: The frequency at which a moving member or a circuit has a maximum output for a given input.

Respirocrit: In medical nanorobotics, the volume-fraction or bloodstream concentration of respirocyte nanorobots, expressed as a percentage.

Respirocyte: In medical nanorobotics, a theorized bloodborne spherical 1-micron (nanorobotic) device having a 1000-atm pressure vessel with active pumping powered by endogenous serum glucose, that serves as a mechanical artificial red blood cell.

Ribonuclease: An enzyme that cuts RNA molecules into smaller pieces.

SAMFET: (self assembled monolayer field effect transistor). Where a few molecules act as FETs, exhibiting both very strong gain, and extraordinarily rapid response. [Mark Ratner & MT 5(2) p. 20]

Science: The process of developing a systematized knowledge of the world through the variation and testing of hypotheses. The pursuit of knowledge and understanding, from the Latin term *scientia*, which means 'knowledge'.

STAB bonding: Tape automated bonding; semiconductor packaging technique that uses a tiny lead-frame to connect circuitry on the surface of the chip to a substrate instead of wire bonds.

Western blotting: Technique by which proteins are separated and immobilized on a paper sheet and then analyzed, usually by means of a labeled antibody.

Bibliography

Bueno, O.: 2004, 'Von Neumann, Self-Reproduction and the Constitution of Nanophenomena', in: D. Baird, A. Nordmann & J. Schummer (eds.), *Discovering the Nanoscale*, Amsterdam: IOS Press, pp. 101-115.

ETC: 2003b, 'No Small Matter II: The Case for a Global Moratorium ' Size Matters!' *Occasional Paper Series* 7(1) [www.etcgroup.org/documents/Occ.Paper_Nanosafety.pdf].

Fogelberg, H. & Glimell, H.: 2003, *Bringing Visibility to the Invisible: Towards A Social Understanding of Nanotechnology*, Göteborg: Göteborg University [www.sts.gu.se/publications/STS_report_6.pdf].

Fogelberg, H.: 2003, 'The Material Culture of Nanotechnology', in: H. Fogelberg & H. Glimell, *Bringing Visibility to the Invisible: Towards A Social Understanding of Nanotechnology*, Göteborg: Göteborg University, pp. 99-114.

Glimell, H.: 2004, 'Grand Visions and Lilliput Politics: Staging the Exploration of the 'Endless Frontier'', in: D. Baird, A. Nordmann & J. Schummer (eds.), *Discovering the Nanoscale*, Amsterdam: IOS Press, pp. 231-246.

Gorman, M.E.; Groves, J.F. & Shrager, J.: 2004, 'Societal Dimensions of Nanotechnology as a Trading Zone: Results from a Pilot Project', in: D. Baird, A. Nordmann & J. Schummer (eds.), *Discovering the Nanoscale*, Amsterdam: IOS Press, pp. 63-73.

Grinbaum, A.: 2004, 'La condition de l'homme moderne et les nanotechnologies', in: G. Nivat (ed.), *Les limites de l´humain. 39èmes Rencontres Internationales de Genève*, L´Age d´Homme, Genève, p. 141.

Gupta, V.K. & Pangannaya, N.B.: 2000, 'Carbon nanotubes: bibliometric analysis of patents', *World Patent Information*, 22, 185-189.

Haruta, M. 1997. *Catalyst surveys of Japan* 1:61 and references therein.

Hessenbruch, A.: 2004, 'Nanotechnology and the Negotiation of Novelty', in: D. Baird, A. Nordmann & J. Schummer (eds.), *Discovering the Nanoscale*, Amsterdam: IOS Press, pp. 135-144.

Huo, Q., D.I. Margolese, U. Ciesla, P. Feng, T.E. Gier, P. Sieger, R. Leon, P.M. Petroff, F. Schuth, and G.D. Stucky. 1994. *Nature* 368:317.

Johansson, M.: 2003, 'Plenty of room at the bottom: Towards an anthropology of nanoscience', *Anthropology Today*, 19 (No. 6), 3-6.

Jounet, C., W.K. Maser, P. Bernier, A. Loiseau, M. Lamy de la Chapelle, S. Lefrant, P. Deniard, R. Lee, and J.E. Fischer. 1997. *Nature* 388:756.

Khushf, G.: 2004, 'A Hierarchical Architecture for Nano-scale Science and Technology: Taking Stock of the Claims About Science Made By Advocates of NBIC Convergence', in: D. Baird, A. Nordmann & J. Schummer (eds.), *Discovering the Nanoscale*, Amsterdam: IOS Press, pp. 21-33.

Krätschmer, W., L.D. Lamb, K. Fostiropoulos, and D.R. Huffman. 1990. *Nature* 347:354.

Kupperman, A., S. Nadimi, S. Oliver, G. Ozin, J. Garcés, and M. Olken. 1993. *Nature* 365:239.

Kuusi, O.; Meyer, M.: 2002, 'Technological generalizations and leitbilder' the anticipation of technological opportunities', *Technological Forecasting & Social Change*, 69, 625-639.

Landon, B.: 2004, 'Less is More: Much Less is Much More: The Insistent Allure of Nanotechnology Narratives in Science Fiction', in: N.K. Hayles (ed.), *Nanoculture: Implications of the New Technoscience*, Bristol, UK: Intellect Books, pp. 131-146.

Laszlo, P.: 2004, 'Is There Life After Partington?', *Hyle: International Journal for Philosophy of Chemistry*, 10(2), 169-178.

Lenhard, J.: 2004, 'Nanoscience and the Janus-Faced Character of Simulations', in: D. Baird, A. Nordmann & J. Schummer (eds.), *Discovering the Nanoscale*, Amsterdam: IOS Press, pp. 93-100.

López, J.: 2004, 'Bridging the Gaps: Science Fiction in Nanotechnology', *Hyle: International Journal for Philosophy of Chemistry*, 10(2), 129-152.

Lösch, A.: 2004, 'Nanomedicine and Space: Discursive Orders of Mediating Innovations', in: D. Baird, A. Nordmann & J. Schummer (eds.), *Discovering the Nanoscale*, Amsterdam: IOS Press, pp. 193-202.

Marshall, K.: 2004, 'Atomizing the Risk Technology', in: N.K. Hayles (ed.), *Nanoculture: Implications of the New Technoscience*, Bristol, UK: Intellect Books, pp. 147-160.

Mehta, M.D.: 2002, 'Nanoscience and Nanotechnology: Assessing the Nature of Innovation in These Fields', *Bulletin of Science, Technology & Society*, 22(4), 269-273.

Mehta, M.D.: 2004, 'From Biotechnology to Nanotechnology: What Can We Learn from Earlier Technologies?, *Bulletin of Science, Technology & Society*, 24, 34-39.

Meyer, M. & Kuusi, O.: 2004, 'Nanotechnology: Generalizations in an Interdisciplinary Field of Science and Technology', *Hyle: International Journal for Philosophy of Chemistry*, 10(2), 153-168.

Mnyusiwalla, A.; Abdallah, S.D.; Singer, P.A.: 2003, 'Mind the gap: science and ethics in nanotechnology', *Nanotechnology*, 14, R9-R13.

Mody, C.C.M.: 2004, 'How Probe Microscopists Became Nanotechnologists', in: D. Baird, A. Nordmann & J. Schummer (eds.), *Discovering the Nanoscale*, Amsterdam: IOS Press, pp. 119-133.

Mody, C.C.M.: 2004, 'Instruments in Training: The Growth of American Probe Microscopy in the 1980s', in: D. Kaiser (ed.), *Pedagogy and the Practice of Science: Producing Physical Scientists, 1800-2000*, Cambridge, MA: MIT Press (forthcoming).

Mody, C.C.M.: 2004, 'Small, but Determined: Technological Determinism in Nanoscience', *Hyle: International Journal for Philosophy of Chemistry*, 10(2), 99-128.

Nordmann, A.: 2004, 'Was ist TechnoWissenschaft? ' Zum Wandel der Wissenschaftskultur am Beispiel von Nanoforschung und Bionik', in: T. Rossmann & C. Tropea (eds.), *Bionik ' Neue Forschungsergebnisse aus Natur-, Ingenieur- und Geisteswissenschaften*, Berlin: Springer, 2004 (forthcoming)

Paschen, H.; Coenen, C.; Fleischer, T.; Grünwald, R.; Oertel, D. & Revermann, C.: 2004, *Nanotechnologie: Forschung, Entwicklung, Anwendung*, Berlin: Springer (*Nanotechnologie, TAB-Arbeitsbericht 92*, Berlin: Büro für Technikfolgen-Abschätzung beim Deutschen Bundestag).

Roberts, J.A.: 2004, 'Deciding the Future of Nanotechnologies: Legal Perspectives on Issues of Democracy and Technology', in: D. Baird, A. Nordmann & J. Schummer (eds.), *Discovering the Nanoscale*, Amsterdam: IOS Press, pp. 247-255.

Robinson, C.: 2004, 'Images in NanoScience/Technology', in: D. Baird, A. Nordmann & J. Schummer (eds.), *Discovering the Nanoscale*, Amsterdam: IOS Press, pp. 165-169.

Index

F

G

H

I

J

K

L

M

N